COMMON CORE ACHIEVE

Mastering Essential Test Readiness Skills

GED® Test Exercise Book

MATHEMATICS

Bothell, WA • Chicago, IL • Columbus, OH • New York, NY

GED®, GED TESTING SERVICE®, and GED PLUS® are registered trademarks owned by American Council on Education ("ACE"). This material is not endorsed or approved by ACE or the GED Testing Service LLC.

MHEonline.com

Copyright © 2015 McGraw-Hill Education

All rights reserved. No part of this publication may be reproduced or distributed in any form or by any means, or stored in a database or retrieval system, without the prior written consent of McGraw-Hill Education, including, but not limited to, network storage or transmission, or broadcast for distance learning.

Send all inquiries to:
McGraw-Hill Education
8787 Orion Place
Columbus, OH 43240

ISBN: 978-0-02-135568-6
MHID: 0-02-135568-1

Printed in the United States of America.

3 4 5 6 7 8 9 RHR 17 16 15 14

Table of Contents

Congratulations! If you are using this book, it means that you are taking a key step toward achieving an important new goal for yourself. You are preparing to take the GED® Test, one of the most important steps in the pathway toward career, educational, and lifelong well-being and success.

Common Core Achieve: Mastering Essential Test Readiness Skills is designed to help you learn or strengthen the skills you will need when you take the GED® Test. The Mathematics Exercise Book provides you with additional practice of the key concepts and core skills and practices required for success on test day and beyond.

How to Use This Book

This book is designed to follow the same lesson structure as the Core Student Module. Each lesson in the Mathematics Exercise Book is broken down into the same sections as the core module, with a page or more devoted to the key concepts covered in each section. Each lesson contains at least one Test-Taking Tip, which helps you prepare for a test by giving you hints such as how to approach certain question types, or strategies such as how to eliminate unnecessary information. At the back of this book, you will find the answer key for each lesson. The answer to each question is provided along with a rationale for why the answer is correct. If you get an answer incorrect, please return to the appropriate lesson and section in either the online or print Core Student Module to review the specific content.

There are two additional resources at the back of this book to further help you. The Mathematical Formulas sheet lists all of the formulas you will need while working on the lessons, and the Calculator Reference Sheet shows you how to use the TI-30XS MultiView™ calculator.

About the GED® Mathematical Reasoning Test

The GED® Mathematical Reasoning Test assesses across two content areas: quantitative and algebraic problem solving, with a breakdown of approximately 45% focusing on quantitative problem solving, and approximately 55% focusing on algebraic problem solving. The test is broken down into two parts: a short calculator-prohibited section, and a longer calculator-allowed section. Multiple item types are used on the test including multiple choice, fill-in-the-blank, drop-down, drag-and-drop, and hotspot. All of the item types may utilize graphs, tables, charts, or other information presented visually.

The GED® Mathematical Reasoning Test assesses across the Webb's Depth of Knowledge spectrum, asking students to answer questions that range from recall questions (DOK 1) to strategic thinking questions (DOK 3). The test assesses approximately 20% of its items at the DOK 1 level (recall), and 80% of its items at the DOK 2 (application of concepts) and DOK 3 (strategic thinking) levels.

On test day, you will be allowed to use the calculator provided onscreen for the calculator portion of the test. You will also be given a formula sheet as well as an erasable note board to write out work by hand. You will not be allowed to bring your own calculator or scrap paper.

Item Types

The GED® test consists of a variety of question types, including multiple choice, fill-in-the-blank, drop-down, drag-and-drop, and hot spot. To prepare you for the GED® test, the Mathematics Exercise Book models those computer-based question types in a print format to help familiarize you with what you will experience on test day.

Multiple-choice Items

The multiple-choice question is the most common type of question you will encounter. Each multiple-choice question will contain 4 answer choices, of which there will be only one correct answer. When encountering a multiple-choice question, look for any possible answers that cannot be correct based on the information given. You may also see extraneous information in the question that is used in the answer choices. Identify and eliminate this information so you can focus on the relevant information to answer the question.

> A square pyramid with a height of 8 centimeters is stacked on a cube. The side of the cube is $2\frac{1}{2}$ times greater than the height of the pyramid. What is the volume of the composite figure to the nearest cubic centimeter?
>
> A. 8,067 cm³
>
> B. 9,067 cm³
>
> C. 11,200 cm³
>
> D. 16,000 cm³

Drop-down Items

The drop-down items are questions that give a drop-down menu within the text with choices to fill in the space to complete the sentence. There can be multiple drop-down items in a text, each with its own set of possible answers. When answering drop-down items, try to eliminate answer choices that are meant as a distraction, including choices with unnecessary information from the text or choices that reuse information from a previous drop-down item. Within this book, these items are simulated by showing an expanded drop down menu from which the correct answer can be selected.

> Company A's cell phone plan charges a monthly fee of $25 plus $2 per GB of data used. Company B's cell phone plan charges a monthly fee of $20 plus $3 per GB of data used. Write an expression showing the total cost of each plan during one month using g GB.
>
> If you use 3 GB of data in a month, Company A costs [1] Select ... ▼ and Company B costs
>
> [2] Select ... ▼ .
>
>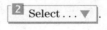
>
> | [1] Select ... ▼ | [2] Select ... ▼ |
> |---|---|
> | A. $29 | A. $29 |
> | B. $31 | B. $31 |
> | C. $34 | C. $34 |
> | D. $38 | D. $38 |

Fill-in-the-blank (FIB) Items

A fill-in-the-blank item has you either complete a sentence by writing in the number, word, or phrase that completes the sentence, or write in the number, word, or phrase that answers a question. If the blank occurs within a sentence, make sure to not only fill in the blank, but to make sure the sentence makes sense, keeping track of verb tenses when writing text and using appropriate units when writing numbers.

> The numbers _____ and _____ would make the following expression undefined.
>
> $20 \div (49 - m^2)$

Drag-and-drop Items

A drag-and-drop activity is an item type in which you are required to drag text or images and drop them in a specific place. Examples of drag-and-drop items include categorizing numbers/expressions/equations, or ordering numbers/expressions/equations from least to greatest according to a particular classification. For a drag-and-drop item, you will be given multiple items that need to be dragged, called draggables. Each draggable will need to classified, categorized, or matched to the appropriate location, or target. Within this book, these items are simulated through writing each draggable in the appropriate target area

List the fruits below in order of unit price, from least to greatest:

4 pounds of | apples | for $5.96

5 pounds of | bananas | for $3.45

6 pounds of | pears | at $7.14

7 pounds of | grapes | at $13.23

8 pounds of | peaches | at $12.72

Least

Greatest

Hotspot Items

A hotspot item consists of an image in which there are interactive spots on a graph or another image. This item type can be used to plot points on graphs, number lines, and dot plots. When engaged with a hotspot item, carefully consider what the directions are asking you to do and then use the interactive image to plot points accordingly to the prompt.

Plot three points from the equation of the line $y = -x$.

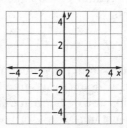

Strategies for Test Day

There are many things you should do to prepare for test day, including studying. Other ways to prepare you for the day of the test include preparing physically, arriving early, and recognizing certain strategies to help you succeed during the test. Some of these strategies are listed below.

- **Prepare physically.** Make sure you are rested both physically and mentally the day of the test. Eating a well-balanced meal will also help you concentrate while taking the text. Staying stress-free as much as possible on the day of the test will make you more likely to stay focused than if you were stressed.

- **Arrive early.** Arrive at the testing center at least 30 minutes before the beginning of the test. Give yourself enough time to get seated and situated in the room. Keep in mind that many testing centers will not admit you if you are late.

- **Think positively.** Studies have shown that a positive attitude can help with success, although studying helps even more.

- **Relax during the test.** Stretching and deep breathing can help you relax and refocus. Try doing this a few times during the test, especially if you feel frustrated, anxious, or confused.

- **Read the test directions carefully.** Make sure you understand what the directions are asking you to do and complete the activity appropriately. If you have any questions about the test, or how to answer a specific item type using the computer, ask before the beginning of the test.

- **Know the time limit for each test.** The Mathematics portion of the test has a time limit of 115 minutes (1 hour 55 minutes). Try to work at a manageable pace. If you have extra time, go back to check your answers and finish any questions you might have skipped.

- **Have a strategy for answering questions.** For each question, read through the question prompt, identifying the important information needed to answer the question. If you need, reread the information provided as well as any answer choices provided.

- **Don't spend a lot of time on difficult questions.** If you are unable to answer a question or are not confident in your answer, you can click on *Flag for Review* in the test window to mark the question and move on to the next question. Answer easier questions first. At the end of the test, you will be able to answer and reviw flagged questions, if time permits.

- **Answer every question on the test.** If you do not know the answer, make your best guess. You will lose points leaving questions unanswered, but making a guess could possibly help you gain points.

Good luck with your studies, and remember: you are here because you have chosen to achieve important and exciting new goals for yourself. Every time you begin working within the materials, keep in mind that the skills you develop in *Common Core Achieve: Mastering Essential Test Readiness Skills* are not just important for passing the GED® Test; they are keys to lifelong success.

This lesson will help you practice ordering and comparing numbers. Use it with core lesson 1.1 *Order Rational Numbers* to reinforce and apply your knowledge.

Key Concept

Rational numbers include whole numbers, fractions, decimals, and their opposites. A number line is a useful math tool for comparing and ordering rational numbers.

Core Skills & Practices

- Use Math Tools Appropriately
- Apply Number Sense

Rational Numbers

We rely on rational numbers to count, measure, and describe situations each day.

Directions: Use the number line below to answer questions 1–2.

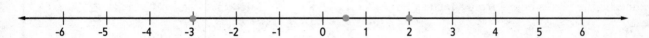

1. Select the rational numbers that are marked on the number line above.

    A. 2, 0.75, 8

    B. -3, 0.5, $\frac{4}{2}$

    C. -6, $\frac{30}{5}$, 3

    D. -0.5, $\frac{-4}{2}$, 3

2. How many whole numbers are labeled on the number line?

    A. 13

    B. 12

    C. 7

    D. 6

Directions: Answer the questions below.

3. Irrational numbers have what property?

    A. They are perfect squares.

    B. They can be expressed as the ratio of a/b, where b is a non-zero integer.

    C. They cannot be expressed as the ratio of two integers.

    D. They are terminating decimals that do not repeat.

4. Compare $\sqrt{2}$ and $\sqrt{1}$; which number is rational and why?

    A. $\sqrt{2}$; it's a perfect square.

    B. $\sqrt{1}$; it's a perfect square.

    C. $\sqrt{2}$; it's an infinite decimal.

    D. $\sqrt{1}$; it's an infinite decimal.

Fractions and Decimals

Fractions and decimals are encountered daily; it's important to understand how to compare and order them.

Directions: Answer the questions below.

5. Convert $\frac{28}{5}$ to a mixed number.

   A. $5\frac{2}{3}$

   B. $7\frac{3}{5}$

   C. $7\frac{2}{3}$

   D. $5\frac{3}{5}$

6. A runner logs her runs using a phone. The table below shows her distances for four days. Write the distances in the boxes below from shortest to longest.

Mon	Tues	Wed	Thurs
$2\frac{6}{10}$	$2\frac{3}{12}$	$2\frac{3}{4}$	$2\frac{3}{8}$

SHORTEST

LONGEST

 Test-Taking Tip

When completing a drag-and-drop activity, read carefully to determine how you should rank, or order, the information provided. Then, select the most logical sequence to correctly complete the question.

7. An incomplete expression is shown. Write the symbol that completes the expression.

0.125 _____ $\frac{1}{6}$

Æ Symbol				⊠
π	f	≥	≤	≠
²	³	|	×	÷
±	∞	√	+	—
(	)	>	<	=
				Insert

8. You are painting the walls of three units in an apartment building. Your supervisor says you will need $5\frac{3}{4}$ gallons of paint for the three units. The amount of paint in each can is written in decimals. Which of these is the same as $5\frac{3}{4}$ gallons?

   A. 0.575 gallon

   B. 5.34 gallons

   C. 5.75 gallons

   D. 53.4 gallons

9. Which of the following numbers is the largest?

   A. 10.02

   B. 10.226

   C. 10.5

   D. 10.23

Absolute Value

Absolute value describes the distance a number is from zero.

Directions: Use the number line below to answer questions 10–11.

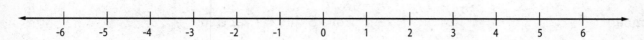

10. Which numbers are a distance of 3 units from the number 2?

   A. −1, 5

   B. −3, 9

   C. −1, 1

   D. −3, 5

11. Find the distance between −6 and 3.

   A. 3

   B. −3

   C. 9

   D. −9

Directions: Answer the questions below.

12. The absolute value of the sum of $-17 + 8$ is _____ .

13. Using the absolute value of each number, order the numbers below from least to greatest.

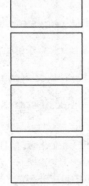

$|-1|$

6

$|-12|$

19

$|-20|$

LEAST
GREATEST

14. Which equation reflects the absolute value of the distance between −4 and −6?

   A. $(-6) - (-4)$

   B. $-6 - 4$

   C. $|-4 - 6|$

   D. $|(-4) - (-6)|$

15. At the warehouse where you work, you are in charge of the inventory. The warehouse has 217 gym bags in stock. You receive an order for 538 bags. How many bags are now on back order?

   A. 755

   B. −755

   C. −321

   D. 321

This lesson will help you practice adding, subtracting, multiplying and dividing rational numbers. Use it with core lesson 1.2 *Apply Number Properties* to reinforce and apply your knowledge.

Key Concept

The least common multiple and greatest common factor of a pair of numbers can be used to solve problems. Awareness of number properties can be helpful in evaluating numerical expressions, although some expressions are undefined.

Core Skills & Practices

- Apply Number Sense Concepts
- Perform Operations

Factors and Multiples

You can use the concepts of factors and multiples to figure out how to break up a number of items into smaller groups or to determine how many items you will need to complete a task.

Directions: Answer the questions below.

1. Complete the tree diagram. Write your answer on each line.

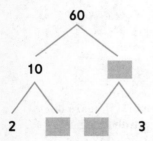

2. Write the prime factorization of 60 in exponent form, by filling in the missing numbers below.

$60 = 2^{\square} \times \square \times \square$

3. There are 20 distance runners and 25 sprinters in a running club. The head of the club wants to organize the club members into equal-sized groups. Each group will have the same number of distance runners and sprinters. How many groups will there be if each group has the greatest possible number of runners?

A. 2

B. 4

C. 5

D. 10

4. A craftsman purchases materials to make dog collars for a pet shop. There are 12 buckles in a pack and 16 straps in a pack. What is the least number of packs of buckles and straps the craftsman should buy so there are no supplies left over?

A. 16 packs of buckles and 12 packs of straps

B. 4 packs of buckles and 3 packs of straps

C. 2 packs of buckles and 1 pack of straps

D. 1 pack of buckles and 1 pack of straps

5. An artist is making a sketch based on a 12-inch by 30-inch poster. She divides the poster into grid squares. What are the greatest size grid squares she can make?

A. 1 square inch

B. 2 square inches

C. 3 square inches

D. 6 square inches

Properties of Numbers

There are certain properties of numbers you can use to help make your calculations easier.

Directions: Answer the questions below.

6. Which expression completes the equation?

   $2(6 + 3) = 2 \times 6 +$ _____

   A. 2×3

   B. $2 + 3$

   C. 2×6

   D. $2 + 6$

7. Which equation illustrates the Commutative Property of Addition?

   A. $x(y + z) = xy + xz$

   B. $x + 0 = x$

   C. $x + (y + z) = (x + y) + z$

   D. $x + y = y + x$

8. Write the property illustrated by each example in the table below.

Associative Property of Multiplication	Commutative Property of Multiplication

Commutative Property of Addition	Distributive Property

Example	Property
$(2 \times 3) \times 4 = 2 \times (3 \times 4)$	
$7 + (10 + 3) = 7 + (3 + 10)$	
$4(a + 6) = 4a + 24$	
$6 \times 7 = 7 \times 6$	

9. The table below shows the numbers of tee shirts sold by three vendors at a concert this weekend. The total number of tee shirts sold by Vendor A is 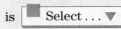 Select . . . ▼ the total number of tee shirts sold by Vendor C.

	Vendor A	Vendor B	Vendor C
Friday	85	101	92
Saturday	101	92	101
Sunday	92	85	85

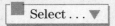 Select . . . ▼

A. greater than

B. less than

C. equal to

D. half

✔ **Test-Taking Tip**

When completing a cloze exercise, read the sentence filling in each possibilty until the sentence makes sense and is mathematically correct.

Order of Operations

When evaluating expressions, the order in which different operations are performed has a direct impact on the final answer, so rules and conventions must be followed.

Directions: Answer the questions below.

10. The numbers _____ and _____ would make the following expression undefined.

$20 \div (49 - m^2)$

11. If 100 is the best possible score on a test and any score over 70 is passing, is a score of $100 - \dfrac{10}{2} \times 20$ a passing test score?

_____

12. Kevin made a mistake while evaluating the expression below.

$2(16 \div (8 - 6)^3 + 5)$
$2(2 - 216 + 5)$
$2(-209)$
-418

What should he do to correct his mistake?

A. Work in order from right to left to multiply and divide, then evaluate the exponent.

B. Perform operations outside the parentheses first, then work from left to right.

C. Work in order from left to right to add and subtract, then multiply and divide.

D. Do the operation in the inner set of parentheses before evaluating the exponent.

13. Marcie buys 3 shirts at $15.00 each. She also buys a $30.00 jacket that is on sale at a $5.00 discount. She uses a $10.00 gift card towards her purchase. Which expression shows how much Marcie spends?

A. $(3 \times 15 + (30 - 5)) - 10$

B. $(3 \times 15 + (30 - 5)) + 10$

C. $(3 + 15 + (30 - 5)) - 10$

D. $(3 + 15 + (30 - 5)) + 10$

14. Which expression equals 21?

A. $\dfrac{5 + 6^2 - 10}{2 + 3}$

B. $\dfrac{(5 + 6)^2 - 10}{2 + 3}$

C. $5 + \dfrac{(6^2 - 10)}{2} + 3$

D. $\dfrac{5(6^2 - 10)}{2 + 3}$

This lesson will help you use exponents and scientific notation to solve real-world problems. Use it with core lesson 1.3 *Compute with Exponents* to reinforce and apply your knowledge.

Key Concept

Exponents can be used to represent and solve problems, such as those involving squares and cubes or scientific notation. You can use the rules of exponents to rewrite and simplify expressions involving exponents.

Core Skills & Practices

- Represent Real-World Problems
- Make Use of Structure

Exponential Notation

Exponential notation is a way to express repeated multiplication. It is useful for computing investments and measuring areas.

Directions: Answer the questions below.

1. Ceramic tile costs $9.50 per square foot, and the installation fee for one room of tile is $75. Which expression can be used to find the total cost of installing ceramic tile in a room that is 14 feet by 14 feet?

   A. $9.50^2 \times 14 + 75$

   B. $9.50^2 \times 14^2 + 75$

   C. $9.50 \times 14^2 + 75^2$

   D. $9.50 \times 14^2 + 75$

2. What is the value of the expression $x^2 + y^3$ for $x = 4$ and $y = 2$? _____

3. The table shows the cost of sod at two different garden shops. Sam purchased his sod from shop A and bought enough sod to cover an area that is 20 feet by 20 feet. Isaiah purchased his sod from shop B and bought enough sod to cover an area that is 16 feet by 16 feet. Matt purchased his sod from shop C and bought enough sod to cover an area of 15 feet by 15 feet. Write and evaluate three expressions, using exponents, to figure out how much each person spent. Then write the names in order from least to greatest to show who spent the most money.

Sam	Isaiah	Matt

Garden Shop	Sod Cost per Square Foot
A	$0.40
B	$0.50
C	$0.60

Least Money

Most Money

Directions: Answer the questions below.

4. Juan is planting flower seeds in 6 large cubic containers and 1 small cubic container. The side length of each large container measures 4 inches, and the side length of the small container measures 3 inches. Which expression shows the number of cubic inches of dirt Juan will need to completely fill all the containers with soil?

   A. $6 \times 4^3 + 3^3$

   B. $6 \times 4^2 + 3^2$

   C. $6 \times 4^3 + 1 + 3^3$

   D. $6 \times 4^2 + 1 + 3^2$

5. The value of the expression $13 \times a^0 + 1$ is

   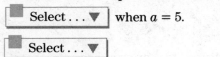 Select . . . ▼ when $a = 5$.

   Select . . . ▼

   A. 1

   B. 5

   C. 13

   D. 14

Rules of Exponents

The rules of exponents help you simplify expressions involving exponents and make them easier to solve.

Directions: Answer the questions below.

6. Which property can be used to simplify the expression $(4a^4)^2$?

   A. Product of Powers Property

   B. Quotient of Powers Property

   C. Power of a Power Property

   D. Distributive Property

7. Which value completes the equation $(t\_\_\_)^4 = t^{-12}$? Write your answer in the line.

8. Write the expressions from least to greatest.

   $12^2 \times 12^5$

   $(12^2)^5$

   $12^5 \div 12^2$

Least
Greatest

9. What is the value of the expression $\frac{4^{-2} \times 4^4}{4^2}$?

   A. 1

   B. 2

   C. 4

   D. 16

10. What is the value of the expression $(2^2 \cdot 3^3)^4$?

    A. 6^{20}

    B. 6^{24}

    C. $2^6 \cdot 3^7$

    D. $2^8 \cdot 3^{12}$

 Test-Taking Tip

You may sometimes see dots in expressions like this on the test. They carry the same meaning as the × symbol and indicate multiplication.

Scientific Notation

Scientific notation simplifies calculations with very small or very large numbers. It is useful for computing long distances, measuring small objects, and modeling populations.

Directions: Answer the questions below.

11. Olivia says 65.2×10^5 is the number 6,520,000 written in scientific notation. Is she correct? If not, explain how she can correct her mistake.

   A. Yes, it is written in scientific notation. No correction needed.

   B. No, it is not written in scientific notation. She needs to move the decimal one more place to the left and leave the exponent as is.

   C. No, it is not written in scientific notation. She needs to leave the decimal as is, and make the exponent 10^6.

   D. No, it is not written in scientific notation. She needs to move the decimal one more place to the left and make the exponent 10^6.

12. What is 0.0000000504 written in scientific notation?

   A. 5.4×10^{-8}

   B. 5.04×10^{-8}

   C. 50.4×10^{-9}

   D. 5.04×10^{9}

13. A biologist can grow 645,300,000 bacteria in one Petri dish. She can grow _____ bacteria in 10 Petri dishes, expressed in scientific notation.

14. What is the sum of the expression $(4.2 \times 10^5) + (6.7 \times 10^3)$ written in scientific notation?

   A. 4.267×10^8

   B. 0.4267×10^6

   C. 4.267×10^5

   D. 42.67×10^4

 Test-Taking Tip

When completing a multiple-choice exercise, read all the answer choices carefully before you select your answer. A decimal point or an exponent number can change an answer completely.

15. A country currently has a population of 3.2×10^7. Its population is expected to double every five years. In what range will the country's population be after 10 years from now?"

   A. 100 million to 120 million

   B. 120 million to 140 million

   C. 140 million to 160 million

   D. 160 million to 180 million

This lesson will help you practice computing with roots using rules of exponents. Use it with core lesson 1.4 *Compute with Roots* to reinforce and apply your knowledge.

Key Concept

Roots, including square roots and cube roots, often appear in real-world problems. Numerical expressions involving roots (often called radicals) can be written using rational exponents and then simplified using the rules of exponents.

Core Skills & Practices

- Represent Real-World Arithmetic Problems
- Attend to Precision

Square Roots and Cube Roots

Just as you can use squares and cubes to find area and volume, you can use square roots and cube roots to find side lengths from an area or a volume.

Directions: Answer the questions below.

1. The _____ root of 27 is 3.

2. A restaurant has been using square plates with a side length of 10 in. The manager decides to switch to plates that will hold 1.5 times as much food. To the nearest tenth of an inch, what must be the side length of the new plates?

   A. 12.2 in.

   B. 13.2 in.

   C. 14.4 in.

   D. 15 in.

3. If a gallon of paint covers 350 ft^2, which of the following is the side length of the largest square floor space that you could paint with one gallon?

   A. 7.0 ft

   B. 7.6 ft

   C. 18.7 ft

   D. 20.0 ft

4. One cubic foot of a swimming pool holds 7.48 gallons of water. Which of the following is the shortest possible side length of a cubical pool that can hold 10,000 gallons?

   A. 10 ft

   B. 11 ft

   C. 12 ft

   D. 13 ft

5. The square root of an integer is either an integer or a(n) _____.

6. If x and y are negative integers, y could be

   A. the square root of x.

   B. the cube root of x.

   C. the fourth root of x.

   D. the sixth root of x.

7. Order the numbers below from least to greatest.

13

$\sqrt{156}$

$\sqrt{140}$

12

$\sqrt{170}$

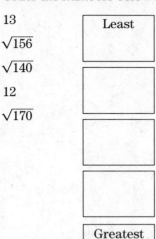

8. Write the following numbers under the appropriate headings.

1, 8, 27, 64, 100, 144

Perfect Square	Perfect Cube

9. The 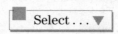 of an integer is sometimes larger than the original number.

[Select . . . ▼]

A. cube root

B. square root

C. fourth root

D. eighth root

10. How many feet of fence are required to enclose a square area of 56.25 ft²?

A. 7.5 ft

B. 28.125 ft

C. 30 ft

D. 32 ft

Radicals and Rational Exponents

The rules of multiplying and dividing powers allow us to work with rational numbers used as exponents.

Directions: Answer the questions below.

11. A square window has a side length of 1 m. The diagonal measures $\sqrt{2}$ m. A fly can walk diagonally across the window in 1 minute. To the nearest tenth of a minute, how long will it take the fly to walk around the perimeter of the window?

A. 0.7 min

B. 1.4 min

C. 2.8 min

D. 4 min

12. Which expression does NOT equal 27?

A. $9^{\frac{3}{2}}$

B. $81^{\frac{3}{4}}$

C. $\sqrt{9^3}$

D. $\sqrt[3]{9^2}$

13. The simplified answer to the expression $\dfrac{2}{\sqrt{2}}$ is

_____.

14. You want to construct a cube-shaped box with a volume of 750 cm³. Which expression shows the area, in square centimeters, of one face of the cube?

A. $750^{\frac{2}{3}}$

B. $750^{\frac{3}{2}}$

C. $6(750^{\frac{1}{3}})$

D. $\dfrac{750^{\frac{2}{3}}}{6}$

15. Simplify $\dfrac{\sqrt{96}}{\sqrt{12}}$.

A. $\dfrac{2}{\sqrt{2}}$

B. $\sqrt{2}$

C. $\sqrt{6}$

D. $2\sqrt{2}$

16. The area of the top of a cube-shaped trash bin is 800 in². To the nearest cubic inch, what is the volume of the trash bin?

A. 4800 in³.

B. 22,627 in³.

C. 2,370,370 in³.

D. 512,000,000 in³.

17. The simplified answer to the expression $\dfrac{\sqrt[3]{108}(\sqrt[3]{16})}{\sqrt{9}}$ is _____.

18. Which of the following expressions is equivalent to $\dfrac{\sqrt{40} \times \sqrt[3]{8}}{\sqrt{2}}$?

A. $\dfrac{2\sqrt{10} \times \sqrt[3]{8}}{\sqrt{2}}$

B. $\dfrac{10^{\frac{1}{2}} \times 8^{\frac{1}{3}}}{2^{\frac{1}{2}}}$

C. $5^{\frac{1}{2}} \times \dfrac{2}{2^{\frac{1}{2}}}$

D. $\sqrt{5} \times \dfrac{\sqrt{2}}{2}$

19. Simplify $\sqrt{\sqrt[6]{2}}$.

A. $\sqrt[12]{2}$

B. $\sqrt[8]{2}$

C. $\sqrt[3]{2}$

D. $2^{\sqrt{6}}$

✓ Test-Taking Tip

Certain operations with roots are easy to confuse. For instance, $x^{\frac{3}{2}}$ is not half of x^3, it is the square root of x^3. The square root of the cube root of 2 is not $2^{\frac{3}{2}}$, but $2^{\frac{1}{6}}$. Think carefully about how root operations work before selecting your answer.

20. The simplified answer to the expression $\dfrac{\sqrt{25}}{\sqrt[3]{125}}$ is _____.

This lesson will help you practice setting up and calculating with ratios, proportions, and scale factors. Use it with core lesson 2.1 *Apply Ratios and Proportions* to reinforce and apply your knowledge.

Key Concept

A ratio, which is often written as a fraction, is a comparison of the relative sizes of two numbers. Operations on ratios follow the same rules as operations on fractions. When two ratios are equivalent, they are called proportional.

Core Skills & Practices

- Compare Unit Rates
- Use Ratio Reasoning

Ratios

Ratios occur throughout your daily routine including miles per hour for speed and cost per pound for fruits and vegetables.

Directions: Answer the questions below.

1. There are 500 boys to 120 girls in a college. Choose an equivalent ratio of boys to girls.

    A. $\frac{50}{9}$

    B. $\frac{50}{3}$

    C. $\frac{25}{6}$

    D. $\frac{25}{12}$

2. In a store, screws are sold for $0.48 a pound. You want to buy 7 pounds for one project and 5 pounds for another project. How much will you spend on screws for both projects?

    A. $0.04

    B. $2.40

    C. $3.36

    D. $5.76

3. On a farm, there are 24 cows, 16 chickens, 20 horses, and 15 pigs. Which ratio of the group is the largest?

    A. Cows : Chickens

    B. Pigs : Cows

    C. Chickens : Horses

    D. Horses : Pigs

4. One car travels 240 miles in 4 hours, while another car travels 275 miles in 5 hours. What is the difference in the rates of speed?

    A. 1 miles/hour

    B. 5 miles/hour

    C. 7 miles/hour

    D. 35 miles/hour

5. List the fruits below in order of unit price, from least to greatest:

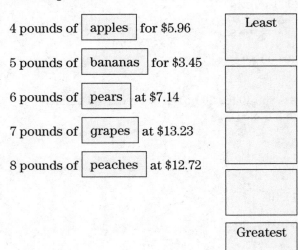

    4 pounds of apples for $5.96

    5 pounds of bananas for $3.45

    6 pounds of pears at $7.14

    7 pounds of grapes at $13.23

    8 pounds of peaches at $12.72

    Least

    Greatest

Proportions

Proportions are evident when changing the amount used in a recipe, reading maps, converting measurements and other applications.

Directions: Answer the questions below.

6. Megan earned $48 working 5 hours Friday afternoon. She plans to work 3 more hours Saturday morning. She made an error when setting up the proportion below to find the amount (t) she will earn Saturday if she is paid at the same rate.

$$\frac{t}{5} = \frac{48}{3}$$

What error did she make? Set up the correct proportion and determine how much would she earn on Saturday.

   A. She switched 3 and 5. She would earn $28.80 on Saturday.

   B. She switched 48 and 5. She would earn $28.80 on Saturday.

   C. She switched 3 and 5. She would earn $80 on Saturday.

   D. She switched 48 and 5. She would earn $80 on Saturday.

7. A company can make 500 screws in 4 seconds. How many screws can the company make in 1 minute?

   A. 125 screws

   B. 2,000 screws

   C. 7,500 screws

   D. 120,000 screws

8. It costs $68.40 for 20 boxes of cereal, how much would 5 boxes cost?

   A. $0.68

   B. $3.42

   C. $17.10

   D. $34.20

9. The exchange rate for U.S. to Chinese currency is 2 dollars to 12 yuan. If you had 883.14 Chinese yuan, you would have _____ U.S. dollars.

 Test-Taking Tip

When writing a proportion, make sure the two equal ratios compare the same items and have the same units.

10. Write a proportion for the situation below, using x as the unknown number of gallons of paint. Then, determine the value of x.

Two gallons of paint cover 800 square feet, and you want to paint 1,800 square feet. How many gallons of paint do you need?

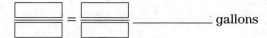 gallons

11. In your household, 4 out of 10 pieces of mail are not read. If you get 2,500 pieces of mail over the course of a year, how much mail is not read?

   A. 10 pieces

   B. 400 pieces

   C. 500 pieces

   D. 1,000 pieces

12. A 1.2 ounce piece of almond candy contains about 230 calories. A 4-ounce piece of almond candy contains about [Select . . . ▼] more calories than the 1.2 ounce candy. Use a proportion to find the answer.

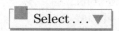

   A. 537

   B. 690

   C. 750

   D. 767

Scale

Scale drawings are used in many applications of building and maps because full-size drawings can be ineffective and impractical.

Directions: Answer the questions below.

13. To find the distance between two nearby cities, Shawn looks at a state map. He measures the map distance between the cities to be almost exactly $2\frac{1}{2}$ inches. The map scale reads 1 inch = 15 miles. Complete and solve the proportion below to find the distance between the cities. Write the missing numbers in the boxes.

$$\frac{15}{\boxed{}} = \frac{d}{\boxed{}}$$

$d = \boxed{}$ miles

14. Jessica is 5.5 feet tall and casts an 8-foot shadow during the day. At the same time, a tree casts a 14-foot shadow. How tall is the tree rounded to the nearest thousandth of a foot?

   A. 3.143 feet

   B. 9.625 feet

   C. 11.500 feet

   D. 20.364 feet

15. Mike's regular pentagonal garden has sides measuring 8 feet. His neighbor builds a similar garden with a scale factor of $\frac{3}{2}$ to Mike's garden. What is the perimeter of his neighbor's garden?

   A. 20 feet

   B. 40 feet

   C. 60 feet

   D. 80 feet

16. A wall in a new house has a height of 17 feet and a width of 19 feet. In a photo of the house, the width of the wall is 0.5 foot. _____ is the scale factor of the photo compared to the actual wall.

Directions: Use the paragraph below to answer questions 17–18.

An isosceles triangle has a base of 6 inches. Each leg is triple the length of the base. A similar isosceles triangle has a base of 10 inches.

17. What is the scale factor of the large triangle to the small triangle?

   A. $\frac{3}{5}$

   B. $\frac{5}{3}$

   C. 3

   D. 5

18. What is the length of one leg of the similar isosceles triangle?

   A. 30 inches

   B. 54 inches

   C. 90 inches

   D. 108 inches

This lesson will help you practice calculating percentages in real-world contexts. Use it with core lesson 2.2 *Calculate Real-World Percentages* to reinforce and apply your knowledge.

Key Concept

A percent is a ratio of a number to 100. In fact, the word *percent* comes from the Latin term *per centum*, meaning "by the hundred," and it is represented by the symbol %. Fractions and decimals are also ratios, and they are related to percents.

Core Skills & Practices

- Use Tools Strategically
- Use Percent

Percent of a Number

Statistical information often appears as a percentage. The percent of a number describes a part of a whole, such as the part of a population that has a certain characteristic.

Directions: Answer the questions below.

1. Shade 63% of the grid that is pictured below.

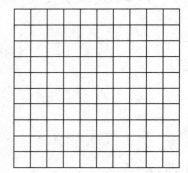

2. A survey was taken by 300 runners. 24% reported that they would rather run barefoot than wear a running shoe. How many runners surveyed prefer wearing a running shoe?

   A. 24 runners

   B. 72 runners

   C. 228 runners

   D. 276 runners

3. A teacher records students' tests in her grade book as percentages. One student got 23 out of 25 correct. What percentage should the teacher record?

   A. 8%

   B. 92%

   C. 11%

   D. 75%

4. Out of 3,000 people surveyed at the beach, 79% of them answered "yes" when asked if they would go into the water above their waist. How many surveyed would go into the water above their waist?

   A. 2,370 people

   B. 237 people

   C. 24 people

   D. 2 people

✓ Test-Taking Tip

Sometimes the correct answer in a word problem involves one more step. Always reread the question after each step to make sure your solution answers the question.

Percent Change

A percent change is a way to compare the difference of an original amount to a new amount.

Directions: Answer the questions below.

5. The Outdoor Swimming Club offers a discounted summer membership if the membership is bought during the winter. The summer membership fee is $126 when purchased during the winter and $180 when purchased during the spring and summer. What is the percent discount during the winter?

A. 70%

B. −54%

C. −30%

D. 30%

6. A golf course offers a membership each year. The first year it opened, the fee was $1,400. Over the next five years the fees were $1,200, $1,250, $1,500, $1,400, and $1450 respectively. What was the percent change between each year? Round to the nearest percent.

year 1 and year 2 _____

year 2 and year 3 _____

year 3 and year 4 _____

year 4 and year 5 _____

year 5 and year 6 _____

7. A clothing store was unable to sell a particular sweater at its original price of $54.99. They decided to give a 20% discount in the hopes that the sweater will sell. If each sweater cost $20 to manufacture, how much profit did the company make selling the sweater to 200 people at the discounted price? Round your answer to the nearest hundred dollars.

A. $2,200.00

B. $4,800.00

C. $8,800.00

D. $10,200.00

8. Which item *saves* you the most money?

A. Original price $12.00; sales discount 30%

B. Original price $43.00; sales discount 5%

C. Original price $36.00; sales discount 15%

D. Original price $25.00; sales discount 20%

9. A company's stock price for 7 consecutive days is listed below.

Day	Stock Price
1	38.00
2	42.50
3	41.00
4	43.50
5	42.00
6	40.25
7	42.75

Use the table above to find the following percent changes between the listed days and fill in the table below. Round to the nearest percent.

	Percent Change
Day 1 to Day 2	
Day 2 to Day 3	
Day 3 to Day 4	
Day 4 to Day 5	
Day 5 to Day 6	
Day 6 to Day 7	
Day 1 to Day 7	

10. The original price of a printer was $212.

If the discounted price is $159.00, then the percent discount is _____.

If the discounted price is $125.08, then the percent discount is _____.

If the discounted price is $106.00, then the percent discount is _____.

If the discounted price is $99.64, then the percent discount is _____.

Simple Interest

Simple interest can be money earned on an investment or money you owe for a loan.

Directions: Answer the questions below.

11. Eric has $2,000 he wants to put into a savings account. Four banks have simple interest accounts that Eric will choose from. Which bank pays the most interest at the end of its term?

    A. Bank A has a simple interest account that pays 3% per year for 4 years.

    B. Bank B has a simple interest account that pays 2.5% per year for 5 years.

    C. Bank C has a simple interest account that pays 1.75% per year for 8 years.

    D. Bank D has a simple interest account that pays 4.2% per year for 3 years.

12. A student borrows $20,000 at a 4.2% simple interest rate per year. The loan period is 15 years.

    The student will pay _____ in simple interest on the loan in one year.

    The student will pay _____ in simple interest on the loan over the life of the loan.

13. Suppose you deposit $10,000 into a bank account that earns simple interest at a rate of 2.99% per year. If you keep your investment in the account and make no more deposits, how much will be in your account after $4\frac{1}{2}$ years?

    A. $1,345.50

    B. $1,196

    C. $11,345.50

    D. $11,196

14. Sean is buying a new car that costs $14,000. He can get a 5-year loan from the car dealership at a simple interest rate of 2.8% per year. He can also get a 3-year loan from his bank at a simple interest rate of 4.99% per year. The better deal for Sean is the _____.

15. Kayla deposits $1,000 into a simple interest savings account on January 1st. She earns a 5% interest rate per year. How many years will it take until she has made $1,000 in interest?

    A. 5 years

    B. 10 years

    C. 15 years

    D. 20 years

16. Chad is looking into buying a car from a used car dealership. He has found 4 different cars that he would be interested in purchasing. However, for each car, he has a different loan option. Each 3-year loan charges simple interest, and Chad would prefer to pay the least amount of interest every month. Which car/loan should he choose?

    A. Car A costs $5,000 with a loan that charges 2.9% monthly interest.

    B. Car B costs $7,500 with a loan that charges 1.8% monthly interest.

    C. Car C costs $11,000 with a loan that charges 1.3% monthly interest.

    D. Car D costs $12,000 with a loan that charges 1.4% monthly interest.

This lesson will help you understand factorials and how to use combinations and permutations to count possibilities in probability situations. Use it with core lesson 2.3 *Use Counting Techniques* to reinforce and apply your knowledge.

Key Concept

Certain events allow for uncertainty. When this occurs, it can be possible to determine the number of possible outcomes by using permutations and combinations.

Core Skills & Practices

• Use Counting Techniques
• Model with Mathematics

Factorials

Factorials show all the different ways a certain number of items can be arranged.

Directions: Answer the questions below.

1. The number 5! equals _____.

2. Jesse is trying to determine the order in which he wants to ride his three favorite roller coasters. He notices that there are 3 choices for the first roller coaster, 2 choices for the second roller coaster, and 2 choices for the last roller coaster. Jesse incorrectly determines the total possible orders by calculating $3 + 2 + 2 = 7$ total possible choices for his ordering. The correct number of possible orders is _____.

3. Which number counts the total number of ways to order 6 people in a line?

   A. 21

   B. 36

   C. 120

   D. 720

4. Clockex sells watches in men's and women's models, with small, medium, and large wristbands. The wristband can be made of cloth, metal, or plastic in a choice of black, green, or ivory. How many different ways can you order a watch?

   A. 18

   B. 27

   C. 54

   D. 81

5. You flip a coin, deal a card from a standard deck, and spin a 4-color spinner. How many different outcomes could there be?

   A. 58

   B. 104

   C. 208

   D. 416

6. Coffee from the Coffee Crew comes in mild, medium, or bold roasts, with or without milk, and with or without sugar. Draw a tree diagram to show how many different ways you can get your coffee at the Coffee Crew. The total number of possible orders from the Coffee Crew is _____.

✔ Test-Taking Tip

If you get confused when working with permutations and combinations, try drawing a tree diagram or making a chart. Then read the context of the question carefully to see if your answer makes sense.

Permutations

Permutations allow us to determine all the different ways a set of items can be ordered when order matters.

Directions: Answer the questions below.

7. Which expression shows the number of permutations of 3 objects taken from 7 objects?

   A. $\dfrac{7!}{(7-3)!}$

   B. $\dfrac{7!}{7!-3!}$

   C. $\dfrac{7!}{3!(7-3)!}$

   D. $\dfrac{7!}{3!(7!-3!)}$

8. How many different parades can be made by lining up 5 available floats?

   A. 25

   B. 120

   C. 125

   D. 3025

9. The number code on your suitcase lock is three digits from 1 to 9, with none repeated, but you've forgotten the number code. If you can try one set of 3 digits every 5 seconds, what is the longest time it could take you to open the lock?

   A. 7 min

   B. 42 min

   C. 1 hr

   D. 1 hr 23 min 15 s

10. When you want to find the number of ordered arrangements of a group of items, you are looking for the number of 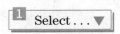 **1** Select . . . ▼ .

    1 Select . . . ▼

    A. combinations

    B. permutations

    C. items

    D. factorials

11. You are directing a historical play. In how many ways can you cast the roles of Washington, Adams, and Jefferson from a pool of 8 actors?

    A. 24

    B. 56

    C. 336

    D. 512

12. You are in charge of planning an assembly. You could choose an opening speaker, a main speaker, and a closing speaker from among 6 candidates in _____ ways.

Combinations

Combinations allow us to specify all the ways to select a certain number of unordered items, like friends at a party.

Directions: Answer the questions below.

13. The program for your assembly is revised. The 3 speakers will be on a panel together. They must still be chosen from 6 candidates. In how many ways could the speakers be chosen from the 6 candidates?

    A. 18

    B. 20

    C. 120

    D. 720

14. Which of these cases involve finding permutations and which involve finding combinations? Write the numerals in the appropriate boxes below.

    I. Electing a president, a vice-president, and a secretary from 7 candidates.

    II. Choosing 5 photographs for an exhibition from 25 entries.

    III. Choosing 3 out of a possible 7 toppings for a dish of ice cream.

    IV. Choosing the order in which to perform a set of 6 songs.

    Permutations   Combinations

15. The table below shows the prices for toppings at a pizzeria.

Toppings: mushroom, onion, anchovy	
1 topping	$6.95
2 toppings	$7.95
3 toppings	$8.95

    What would be the cost to order all of the available offerings of this pizzeria assuming each pizza must have at least one topping?

    A. $53.65

    B. $71.55

    C. $77.50

    D. $95.40

16. The difference between finding permutations and finding combinations is that when you are finding combinations, the _____ of the items does not matter.

17. Keena needs to select 2 friends from a group of 5 friends to invite to have dinner at her house. "Now I have to sort through 20 possible combinations," she incorrectly says. The correct number of combinations is _____.

18. To find the number of combinations of 3 items out of 27, you could find the number of permutations of 3 out of 27 and then divide by what factor?

    A. 2

    B. 3

    C. 6

    D. 9

This lesson will help you understand how to recognize and calculate real-world probabilities. Use it with core lesson 2.4 *Determine Probability* to reinforce and apply your knowledge.

Key Concept

The probability of a chance event uses a number between 0 and 1 to describe the likelihood that the event will occur. You can use the number of total and favorable outcomes of an event to determine the probabilities of simple or compound events.

Core Skills & Practices

• Determine Probabilities

Probability of Simple Events

We make decisions every day based on the likelihood of specific outcomes.

Directions: Answer the questions below.

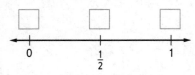

1. Write the letter for each probability on the diagram.

   A. when flipping a coin, the probability of tails showing

   B. on Tuesday, the probability that the next day is Wednesday

   C. on Friday, the probability that the next day is Monday

2. Janine and her two children are attending Family Fun Night at the children's school. A drawing for a door prize will be held at 8:00 P.M. for all attending. If 390 people attend, the probability that someone in Janine's family will win the door prize is _____. Write your answer as a fraction in simplest terms.

3. A company's uniforms consist of different colored shirts and pants. The shirts can be red, blue, or green. The pants can be tan or black. The tree diagram shows the possible color combinations. Which is the probability of randomly choosing a uniform that has a green shirt and black pants?

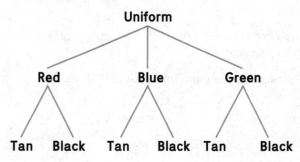

   A. $\frac{1}{9}$

   B. $\frac{2}{9}$

   C. $\frac{1}{6}$

   D. $\frac{4}{6}$

4. Elena thinks of a number from 1 to 50, and Sal tries to guess the number. Suppose Sal guesses the number 28. Sal's guess is more likely to be too _____ (high or low).

5. A bag contains a red marble, a blue marble, a yellow marble, and a green marble. The probability of drawing a red marble is $\frac{1}{4}$. Its

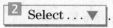 is the probability of drawing

a blue, yellow, or green marble, which is

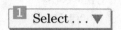.

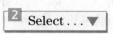

A. supplement

B. complement

C. independent event

D. dependent event

2 Select . . . ▼

A. $\frac{1}{4}$

B. $\frac{1}{2}$

C. $\frac{3}{4}$

D. 1

6. If you roll a cube (numbered 1 through 6) 30 times, which of the following is most likely to occur?

A. rolling a 5 thirty times

B. rolling a 1, 2, or 3 three times

C. rolling a 2 or 4 two times

D. rolling a 3 five times

7. The table below shows the last 25 cell-phone cover purchases at Cell Phone Hut. The best prediction for the number of red covers sold of the next 100 sales is 40. Complete the table.

Probability of Compound Events

Often events happen in conjunction with other events, so the probability that an event occurs may sometimes depend on prior events.

Directions: Answer the questions below.

8. A spinner has 6 equal sections. Two sections are red, 3 sections are blue, and 1 section is green. What is the probability of spinning a red section twice?

A. $\frac{1}{9}$

B. $\frac{2}{6}$

C. $\frac{4}{6}$

D. $\frac{2}{3}$

9. A bag has 5 red marbles and 4 white marbles. Aaron draws a marble, sets it aside, and draws a second marble. What is the probability that Aaron will draw a white marble on both draws?

A. $\frac{1}{6}$

B. $\frac{3}{8}$

C. $\frac{4}{9}$

D. $\frac{5}{6}$

10. Allen chooses one of the cards randomly and then flips the coin. What is the probability that Allen chooses a King and flips a head?

   A. $\frac{1}{8}$

   B. $\frac{1}{4}$

   C. $\frac{2}{6}$

   D. $\frac{1}{2}$

11. Emma chooses one of the cards randomly, sets it aside, and chooses a second card. What is the probability that Emma will choose a Queen for both cards?

   A. $\frac{1}{2}$

   B. $\frac{1}{3}$

   C. $\frac{1}{6}$

   D. $\frac{1}{12}$

 Test-Taking Tip

Make sure you understand the scenario of each question before choosing an answer. Example: whether replacing or setting aside a card makes the second event independent, or dependent on the first event.

Directions: Use the tree diagram to answer questions 12–13.

Peter has a bag of blocks. There is one block for each letter of the alphabet. The tree diagram shows the probabilities of drawing two blocks without replacing the first block after it is drawn.

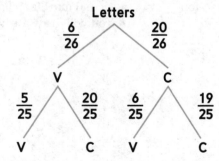

V: vowels
C: consonants

Letters

$\frac{6}{26}$ $\frac{20}{26}$

V   C

$\frac{5}{25}$ $\frac{20}{25}$ $\frac{6}{25}$ $\frac{19}{25}$

V   C   V   C

12. Drawing the second block is a(n) _____ event.

13. What is the probability of drawing a vowel twice?

   A. $\frac{3}{65}$

   B. $\frac{12}{65}$

   C. $\frac{6}{25}$

   D. $\frac{20}{25}$

This lesson will help you use linear expressions to represent problems with one variable. Use it with core lesson 3.1 *Evaluate Linear Expressions* to reinforce and apply your knowledge.

Key Concept

There are a lot of unknowns around us. In math we do not always know the total we are solving for or the values we are calculating. These are expressions. Evaluating linear expressions means substituting values (numbers) for variables (letters).

Core Skills & Practices

- Perform Operations
- Evaluate Expressions

Algebraic Expressions

An algebraic expression can be used to represent the cost of a product, the missing side of a polygon, or the height of an object at a specific time.

Directions: Answer the questions below.

1. A wedding hall charges $10 per person for food and $4 per person for beverages. The rental fee for the wedding hall is $350. Using p = people, which of the following represents the total cost for the wedding hall?

    A. $350 + 10p - 4p$

    B. $350 - 10p - 4p$

    C. $350 + 10p + 4p$

    D. $350 - 10p + 4p$

2. An airplane is at 50,000 feet when it starts to descend to the ground at the rate of 10 feet per second. Using t = time in seconds, choose the expression that represents the plane's descent.

    A. $50,000 - 10t$

    B. $50,000 + 10t$

    C. $10t - 50,000$

    D. $50,000 \times 10t$

3. Let w = width of a garden. The length of the rectangular garden is four times its width, minus 5. What is an expression for the perimeter of the garden?

    A. $4w - 5 + w$

    B. $2(4w - 5) + 2w$

    C. $(2 \times 4w) - 5 + 2w$

    D. $2(4w - 5) + w$

✓ Test-Taking Tip

When reading a problem involving algebraic expressions, find key words in the problem. Words like *plus*, *minus*, *decreasing*, and *increasing* can help you determine the operation(s) needed to write the expression.

4. Robert works at a store during the holidays. He makes $12 per hour for a 40-hour work week and earns 1.5 times as much per hour of overtime. Which expression represents the amount Robert makes in a week (use h = hours of overtime)?

   A. $12(40)$

   B. $12(40) + 1.5h$

   C. $12(40) + 12h$

   D. $12(40) + 12(1.5)h$

5. A brand new truck costs $45,000. The truck depreciates in value by $5,000 as soon as it is driven off the lot, and it loses $300 per month in value after that. Using t = time in months, an expression for the value of the truck over time is _____.

6. Cole cuts a pizza into 12 equal slices. Cole takes n slices of the pizza and shares the remaining slices equally among 5 friends. Which expression tells the number of slices each friend gets?

   A. $\dfrac{(n-12)}{5}$

   B. $\dfrac{(n-5)}{12}$

   C. $\dfrac{(12-n)}{5}$

   D. $\dfrac{(5-n)}{12}$

Linear Expressions

A linear expression is a type of algebraic expression in which terms have no more than two variables.

Directions: Answer the questions below.

7. Put the steps in simplifying the expression $(7x - 8) + 2(x - 5)$ in order.

   $7x + 2x - 10 - 8 = 9x - 18$

   $(7x - 8) + 2(x - 5) = (7x - 8) + 2x - 10$

   $(7x - 8) + 2x - 10 = 7x + 2x - 10 - 8$

First Step

Last Step

8. Choose the first step below that would occur during the simplification of the expression $(-3x + 10) - 4(x - 3)$.

   A. $(-3x + 10) - 4x - 12$

   B. $-7x - 2$

   C. $(-3x + 10) - 4x + 12$

   D. $-7x - 7$

9. Simplify the expression $2(x + 3) - 5(2x + 7) + (6x + 2)$. Write the missing term on the line.

   $- 2x -$ _____

10. Simplify the expression $-5(-3x + 2)$.

    A. $15x - 10$

    B. $-15x + 10$

    C. $-15x + 2$

    D. $15x + 2$

11. Simplify the expression $-2(-4x + 7) - 3(2x - 5) + 4(2x + 6)$. Write the missing number on the line.

_____ $x + 25$

12. Cassandra simplified the expression below.

$(9y - 20) - (13y + 1)$
$9y - 20 - 13y + 1$
$9y - 13y - 20 + 1$
$-4y - 19$

Which of the following describes Cassandra's error?

A. She did not combine like terms into one expression.

B. She did not add the whole numbers correctly.

C. She did not rearrange the expression so like terms are near each other.

D. She did not multiply the 1 by -1 when distributing the coefficients.

13. Look at the expression below. How would the simplified form of the expression change if the coefficient of the last term is positive instead of negative?

$(4x - 17) + (-8x + 9) - 2(x - 14)$

A. It would be $-2x - 36$ instead of $-14x - 36$.

B. It would be $-4x - 20$ instead of $-10x + 54$.

C. It would be $-2x - 36$ instead of $-6x + 20$.

D. It would be $-4x - 54$ instead of $-14x - 2$.

Evaluating Linear Expressions

Linear expressions are evaluated when a value is substituted for the unknown. An example of this is finding the profit of a business for a specific number of items sold.

Directions: Answer the questions below.

14. Evaluate the expression $3x + \frac{3}{2}y$ when $x = -4$ and $y = 2$.

_____

15. Evaluate the following expressions using $x = 5$ and $y = -3$. Then write the expressions into the appropriate categories below.

$2x - 3y \qquad x - y \qquad 2x + y \qquad 3x + y$

Less than 8

Equal to 8

Greater than 8

16. Company A's cell phone plan charges a monthly fee of $25 plus $2 per GB of data used. Company B's cell phone plan charges a monthly fee of $20 plus $3 per GB of data used. Write an expression showing the total cost of each plan during one month using g GB.

If you use 3 GB of data in a month, Company A costs 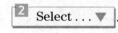 and Company B costs 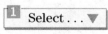.

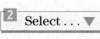

A. $29

B. $31

C. $34

D. $38

A. $29

B. $31

C. $34

D. $38

This lesson will help you solve one- and two-step linear equations using inverse operations. Use it with core lesson 3.2 *Solve Linear Equations* to reinforce and apply your knowledge.

Key Concept	Core Skills & Practices
You can solve an equation by performing inverse operations on both sides of the equation. The solution can be checked using substitution.	• Solve Simple Equations by Inspection • Solve Linear Equations

One-Step Equations

When solving equations, the goal is to find the unknown value. We can use equations to find unknown amounts in many real-world situations.

Directions: Complete each statement.

1. The solution of "A number n tripled is 72" is

   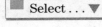 Select . . . ▼ .

   Select . . . ▼

   A. $n = 9$

   B. $n = 24$

   C. $n = 36$

   D. $n = 216$

2. _____ is the equation that represents "8 less than x is 31."

Directions: Answer the questions below.

3. There are 150 people interested in playing in a softball league. The sports director is going to divide all of the people into s teams. Which expression represents the number of players on each team?

   A. $150 \times s$

   B. $150 - s$

   C. $s \div 15$

   D. $150 \div s$

4. On Thursday, a landscaper mowed 6 lawns. After paying $18 for gas, the landscaper had $90 left. Which equation can be used to find the amount of money the landscaper earned on Thursday?

   A. $90 = n + 18$

   B. $90 = n - 18$

   C. $90 = (n + 18) \div 6$

   D. $90 = (n - 18) \div 6$

 Test-Taking Tip

When writing equations from verbal descriptions, look for key words that indicate the operations being performed.

Directions: Answer the questions below.

5. What is the solution of $-7 = w - 10$?

   A. $w = 3$

   B. $w = -3$

   C. $w = 17$

   D. $w = -17$

6. What operation should you use to solve $8 = -56 + x$?

   A. addition

   B. subtraction

   C. multiplication

   D. division

7. Which of the following solutions is true for $-12h = 24$?

   A. $h = 2$

   B. $h = -2$

   C. $h = \frac{1}{2}$

   D. $h = -\frac{1}{2}$

Multi-Step Equations

You can use multi-step equations for more complex calculations involving more than one operation.

Directions: Answer the questions below.

8. What is the solution of "5 plus the product of 4 and z is equal to 49?"

   A. $z = -40$

   B. $z = -11$

   C. $z = 11$

   D. $z = 40$

9. What is the solution of $27 - 6x = -33$?

   A. $x = -1$

   B. $x = 1$

   C. $x = 10$

   D. $x = -10$

10. The amount of money b that Bob has saved and the amount j that Jody has saved are related by the equation $2b + 3 = j$. If Jody has saved $101, how much money has Bob saved?

   A. $49

   B. $52

   C. $202

   D. $205

Directions: Answer the questions below.

11. If $-7(m + 4) = 14$, what is the value of $9m$?

A. -6

B. 6

C. -54

D. 54

12. The difference of a number and 5 is multiplied by -9, and the result is 81. What is the number?

A. -4

B. 4

C. -9

D. 9

13. Keondre ate breakfast at the same café 3 times during the past month. Each time, he ordered the same breakfast and left a \$2 tip. He spent a total of \$39. The cost of each breakfast before tipping is _____.

14. What is the solution of $-18 = -3n + 30 - 9n$?

A. $n = -30$

B. $n = 30$

C. $n = -4$

D. $n = 4$

15. The value of n that makes the equation $12 + n = 7n + 2(n - 3)$ true is _____.

16. If $5y - 7 = 48$, then what does $6y + 4$ equal?

A. 70

B. 11

C. 66

D. 15

17. Write each equation in the box for which the solution is listed.

| $20x + 35 = 15x$ | $-5x + 9 = -2x - 30$ |

| $x - 12 = 3(x - 10) - 8$ | $-x + 13 = 4(x + 12)$ |

| $5x - 1 = -9(x + 11)$ | $-7(8 - x) = 3x - 4$ |

$x = 13$	$x = -7$

18. What is the solution to the equation $12(s - 4) = 5(2s - 1) - 5$?

A. 25

B. 19

C. 37

D. 40

19. Johann loves to build model cars. He currently has 15 model cars that he's made, but he wants to buy more. He is willing to sell some of his model cars so that he can buy an expensive model car, selling for \$52. If he already has \$8 and can sell each of his model cars for \$4, how many model cars will he need to sell to afford the expensive set?

A. 4

B. 8

C. 11

D. 15

This lesson will help you represent real-world problems with linear inequalities and solve algebraically or graphically on a number line. Use it with core lesson 3.3 *Solve Linear Inequalities* to reinforce and apply your knowledge.

Key Concept

Solving linear inequalities is very similar to solving linear equations, except the solution to a linear inequality will include a range of values, called the solution set. The solution set can be graphed on a number line.

Core Skills & Practices

• Represent Real-World Problems
• Solve Inequalities

Inequalities

Inequalities are evident when there is a minimum value or maximum value for an expression.

Directions: Answer the questions below.

1. Which of the following represents the solution to the inequality $x \le -4$?

A.
    -4

B.
    -4

C.
    -4

D.
    -4

2. An admissions officer registers 312 girls in a high school. So far, 71 boys have been registered. If b is the remaining number of boys registered in the school, write an inequality for when there would be more boys than girls in the school.

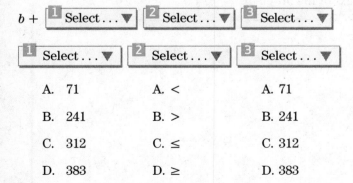

$b +$ ① Select...▼ ② Select...▼ ③ Select...▼

① Select...▼ ② Select...▼ ③ Select...▼

A. 71	A. <	A. 71
B. 241	B. >	B. 241
C. 312	C. ≤	C. 312
D. 383	D. ≥	D. 383

3. Three times the sum of a number and seven is greater than or equal to half of the number. Which of the following inequalities represents this situation?

A. $3x + 7 \ge \frac{1}{2}x$

B. $3x + 7 > \frac{1}{2}x$

C. $3(x + 7) \ge \frac{1}{2}x$

D. $3(x + 7) > \frac{1}{2}x$

4. At a school bake sale, students sell each item for $8.00. The students must raise more than $2400 for a school field trip. Which inequality shows how many items the students must sell?

A.
    300

B.
    300

C.
    300

D.
    300

One-Step Inequalities

One-step inequalities are inequalities that only require one operation to find the solution set.

Directions: Answer the questions below.

5. Which of the following is not a solution to the linear inequality $-3x < 8$.

    A. $x > -\frac{8}{3}$

    B. $x > -2\frac{2}{3}$

    C. $x > -2.666\ldots$

    D. $x > -\frac{3}{8}$

6. Which graph represents the solution to the inequality $x + 12 < 9$?

    A. -3

    B. -21

    C. -21

    D. -3

7. Which of the following inequalities has -8 as a solution? Select all that apply.

    A. $4x \geq -24$

    B. $\frac{x}{2} \leq -3$

    C. $x - 10 > -20$

    D. $x + 16 \leq 8$

8. Which graph represents the solution to the inequality $\frac{x}{-7} > 1$?

    A. 8

    B. -7

    C. -7

    D. 8

9. You need to make at least $1000 profit from selling handmade jewelry and your initial expenses were $400. Which graph represents how many dollars worth of jewelry you need to sell to meet your goal?

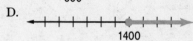

    A. 1400

    B. 600

    C. 600

    D. 1400

 Test-Taking Tip

When working with inequalities, re-read the problem to make sure your answer makes sense within the context of the problem.

Multi-Step Inequalities

Multi-step inequalities have more than one operation within them. Most real-life scenarios that involve inequalities will have multiple operations.

Directions: Answer the questions below.

10. You are ordering pizzas for your youth group with a total of $120 to spend. Each pizza costs $7.75 and you plan to tip 10% for each pizza. You also have a coupon for $10 off. Using p to represent the number of pizzas, write and solve an inequality for this problem. Write your answer on the line.

$p \leq$ _____

11. Solve the inequality $7(4 - a) + 1 < 1 - 4(a + 5)$.

  A. $a < -3$

  B. $a > -3$

  C. $a < 16$

  D. $a > 16$

12. Solve the inequality $\frac{3}{4}x - 8 \geq \frac{2}{3}x - 6$.

  A. $x \geq -24$

  B. $x \geq 24$

  C. $x \leq -\frac{1}{6}$

  D. $x \geq -\frac{1}{6}$

13. Tom scored 91, 74, 83, and 86 on his first four math tests in class. He needs to have an average score of 85 to earn a B in the class, and he has one more test to take. Write and solve an inequality to determine the possible scores he can receive on the test to earn a B for the class.

$x \geq$ _____

14. In order to solve the inequality $5(y + 1) - y \geq -4y + 7$, you should first 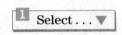 and then simplify both sides of the inequality. Next, getting the variable terms to one side of the inequality and the rest to the other side simplifies to $\boxed{2}$ Select... ▼ before dividing to solve the inequality.

$\boxed{1}$ Select... ▼

  A. add y to both sides of the inequality

  B. subtract 7 from both sides of the inequality

  C. add $4y$ to both sides of the inequality

  D. distribute the 5 through the parenthesis

$\boxed{2}$ Select... ▼

  A. $8y \geq 2$

  B. $8y \geq 7$

  C. $9y \geq 2$

  D. $9y \geq 7$

15. Each inequality below can be simplified to one of the inequalities listed in the bins below. Write each inequality in the correct bin.

$-3y + 5 > -6y + 9$	$-y + 12 < 2(y + 4)$
$7y < y - 8$	$2y > 11(y + 1) + 1$
$-1 < 3y - 5$	$4 + 3(y - 1) < -3$

$y > \frac{4}{3}$	$y < -\frac{4}{3}$

This lesson will help you represent problem situations as expressions, equations, and inequalities and solve the problems. Use it with core lesson 3.4 *Use Expressions, Equations, and Inequalities to Solve Real-World Problems* to reinforce and apply your knowledge.

Key Concept

Real-world problems can be translated into algebraic expressions, equations, and inequalities. Mathematical methods can then be used to find real-world solutions.

Core Skills & Practices

- Evaluate Expressions
- Solve Real-World Problems

Expressions and Equations

We use expressions and equations to model many real-world scenarios.

Directions: Answer the questions below.

1. Nick drives a taxi cab. He charges a flat fee of $4 plus $0.25 per mile. Which expression models the total charge of a ride for m miles?

   A. $m(4 + 0.25)$

   B. $4m + 0.25$

   C. $(4 + m)0.25$

   D. $4 + 0.25m$

2. Freddy gets paid $25 for each lawn that he mows. He would like to buy a bicycle for $110 and has already saved $35. He wants to find the number of lawns he needs to mow to have enough money to buy the bicycle. Which equation models the situation?

   A. $25n = 110 - 35$

   B. $25n - 35 = 110$

   C. $25n - 110 = 35$

   D. $35n + 25 = 110$

3. On a recent trip, Jenny drove 165 miles. She stopped at a rest area for 15 minutes. The entire trip took 3.25 hours. An equation to determine the average speed (s, in miles per hour) that she drove during her trip is _____.

4. Julie uses the equation below to model a situation.

   $$150 - (8h) = 54$$

   Which situation matches the equation?

   A. Julie charges $150 for a cake and $8 for each hour she spends decorating it. She gives the customer a $54 discount.

   B. Julie has $150 to pay for cleaning services. She pays $8 for each hour of cleaning. She has $54 left afterwards.

   C. Julie has $54 to pay someone to paint her living room. She pays $8 for each hour of painting. She has $150 left afterwards.

   D. Julie buys 8 lamps for a total of $150 dollars. She sells some for a profit of $54.

5. Juan can run 1 mile in 12 minutes. Before running, he spends 5 minutes warming up. After running, he spends 10 minutes cooling down. Which expression models the total amount of time he spends running, including his warm–up and cool-down, if he runs m miles?

   A. $5m + 12 + 10$

   B. $12 + m(5 + 10)$

   C. $5 + 1 + 12m + 10$

   D. $5 + 12m + 10$

6. Use the equation $I = Prt$ where I is the amount of interest charged, P is the initial amount borrowed, r is the annual interest rate, and t is the amount of time the money is borrowed.

   Amber, Bailey, and Curt each borrow money from the bank. Amber borrows $3,000 for 3 years at a 6% interest rate. Bailey borrows $1,900 for 6 years at a 5% interest rate. Curt borrows $2,000 for 4 years at a 5% interest rate.

   Write the names in order of the amount of interest charged, from least to greatest.

Amber	Bailey	Curt

Least

Greatest

7. Marcella wants to buy a new treadmill. The cost of the treadmill is $760 plus 5% sales tax. The store will charge a $75 delivery fee. The total amount Marcella will spend on the new treadmill is _____.

8. Kevin bought 4 pounds of apples and 3 pounds of pears at the grocery store. A pound of apples costs $2. Kevin spent a total of $15.50 on the fruit. How much did the pears cost per pound?

   A. $2.38/lb

   B. $2.50/lb

   C. $3.83/lb

   D. $7.83/lb

✔ Test-Taking Tip

When completing ordering activities, make sure you identify whether you are ordering from least to greatest or greatest to least, so that your answer follows the correct order.

Inequalities

We use inequalities to model situations when expressions may or may not be equal.

Directions: Write the phrases below in the appropriate box to match the symbol given.

more than   larger than   fewer than   no less than
minimum   less than   greater than   smaller than
less than or equal to greater than or equal to no more than   maximum

<	≤	>	≥
9.	10.	11.	12.

13. Edward has $1,000 to spend on new carpet in his house. It costs $7 per square foot of carpet. How much carpet can he buy and still have $300 left for the installation fee?

A. no more than 100 square feet

B. no less than 100 square feet

C. more than 700 square feet

D. no less than 700 square feet

14. Hannah's car holds 16 gallons of gas. It can travel up to 32 miles on a gallon of gas. How many miles can Hannah drive on one tank of gas?

A. no more than 2 miles

B. fewer than 2 miles

C. fewer than 512 miles

D. at most 512 miles

15. Maddie has $450 in her bank account. She is going to visit her grandparents in California for a month and she must use her own money to pay for souvenirs and gifts while she is there. However, when she returns home she wants to have enough money left in her account to buy a new pair of boots that cost $230. Maddie can spend _____ each week and still be able to buy the boots.

16. Tara is planning a party. She has a budget of $60. She plans on spending $15 on decorations. She also plans on buying pizzas that each cost $8. Which inequality can be used to determine the number of pizzas that Tara will be able to buy?

A. $8p + 15 \leq 60$

B. $60 + 8p \leq 15$

C. $60 < 8p + 15$

D. $60 > 8p - 15$

17. Jake provides computer services. He charges a flat fee of $85, plus $30 per hour. Which inequality below can be used to find the number of hours he needs to work on a project to earn at least $500?

A. $30 + 85x > 500$

B. $85 + 30x < 500$

C. $85 + 30x \geq 500$

D. $30 + 85x \leq 500$

18. An amusement park has a minimum height requirement of 54 inches to ride a roller coaster. Using h to represent a person's height in inches, the inequality that represents the situation is _____.

19. Jolene has $95 to spend on a night out. Dinner costs $30 and a movie costs $12. The maximum amount Jolene can spend on shopping and still have enough for a $7 cab ride home is _____.

20. Albert earns a base salary of $32,000 a year plus 10% in commission on his total amount in sales. Albert wants to earn more than $50,000 this year. Which expression can be used to find the total amount of sales he needs to make?

A. $50,000 + 10t < 32,000$

B. $32,000 + 0.1t \geq 50,000$

C. $50,000 + 10t \leq 32,000$

D. $32,000 + 0.1t > 50,000$

This lesson will help you represent real-world problems with polynomials and evaluate polynomials for specific values. Use it with core lesson 4.1 *Evaluate Polynomials* to reinforce and apply your knowledge.

Key Concept

Polynomials are special types of variable expressions with one or more terms. Each term has a variable raised to a whole number exponent or is a constant.

Core Skills & Practices

- Use Math Tools Appropriately
- Evaluate Expressions

Identifying Polynomials

Polynomials are algebraic expressions that are a collection of constants, variables, and exponents. Terms that have identical variables with the same exponents are called like terms.

Directions: Use the polynomial shown below to answer questions 1–2.

$$4x^2 + 6x - 3$$

1. Which of the following describes the specific type of polynomial that is shown?

A. monomial

B. binomial

C. trinomial

D. quadrinomial

2. What is the degree of the polynomial?

A. 0

B. 1

C. 2

D. 3

Directions: Answer the questions below.

3. Which of the following binomials is not written in standard form?

A. $9k^2 + 2k$

B. $12 - z$

C. $y^3 + y^2$

D. $2x^2 - 5x$

4. The value of the greatest exponent in the polynomial is the _____ of the polynomial.

5. After simplifying, the degree of the polynomial $(2x - 1)(3x^2 + 5)$ is _____.

6. Which of the following polynomials simplifies to $2x^3 - 4x^2 - x + 2$?

A. $-x^3 + x^2 - 2x + 4 - 2 + x + 5x^2 + 3x^3$

B. $-x^3 + x^2 - 2x + 4 - 2 + x - 5x^2 + 3x^3$

C. $x^3 - x^2 - 2x + 4 - 2 + x - 5x^2 - 3x^3$

D. $x^3 + x^2 - 2x + 4 - 2 + x - 5x^2 + 3x^3$

Evaluating Polynomials

In order to evaluate a polynomial expression for a given value, simply substitute the value into the variable in the expression and simplify the expression completely using order of operations.

Directions: Answer the questions below.

7. Evaluate the polynomial $3x^2 - 2x - 1$ for the following values of x, and order from least to greatest.

$x = -2$

$x = -1$

$x = 0$

$x = 1$

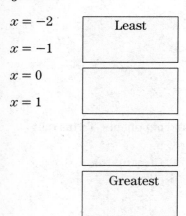

8. The value of the polynomial $5x^3 + x^2 - 10$ when $x = -3$ is _____.

9. The height of a ball being thrown from a cliff that is 120 feet tall is modeled by the polynomial $-16t^2 + 32t + 120$, where t is the number of seconds since the ball was thrown. What is the height of the ball when $t = 2$ seconds?

A. 16 ft

B. 120 ft

C. 136 ft

D. 168 ft

10. If $x = 5$, what is the area of the square shown below?

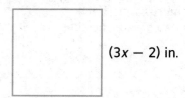

A. 13 in.^{2}

B. 26 in.^{2}

C. 52 in.^{2}

D. 169 in.^{2}

11. The sales of a pool company vary throughout the year and are modeled by the polynomial $-3m^2 + 39m - 48$, where m is the number of the month of the year. How many pools do they typically sell in month 5 of the year, when $m = 5$?

A. 72

B. 120

C. 132

D. 222

 Test-Taking Tip

When evaluating a polynomial expression for a given value, substitute the value into the expression for the variable by first replacing the variable with a set of parentheses and then putting the value into the parentheses. Then, follow the order of operations to simplify the expression completely.

Operations with Polynomials

When working with polynomials, it is often necessary to perform the operations addition, subtraction, and multiplication. So, it is necessary to know and understand the rules for each of these operations when dealing with polynomial expressions.

Directions: Answer the questions below.

12. What is the sum of the polynomials $(x^2 - 5x - 3) + (-x^3 + x - 2)$? Write the answers on the lines.

_____x^3 + _____x^2 - _____x - _____

13. What is the product of $(x - 2)(2x^2 - x + 3)$?

A. $2x^3 - 6x^2 + x - 6$

B. $2x^3 + 3x^2 + x - 6$

C. $2x^3 - 5x^2 + x - 6$

D. $2x^3 - 5x^2 + 5x - 6$

14. The opposite polynomial simply reverses the

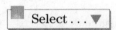 Select . . . ▼ of each term of the original polynomial.

Select . . . ▼

A. signs

B. exponents

C. powers

D. variables

15. The first step for simplifying the polynomial $(3x^3 + 4x^2 - x + 2) - (x^2 + 5x - 2)$ is finding the opposite of the second polynomial before you combine like terms. Which shows the opposite of the polynomial $(x^2 + 5x - 2)$?

A. $x^2 + 5x - 2$

B. $x^2 - 5x - 2$

C. $-x^2 - 5x + 2$

D. $-x^2 - 5x - 2$

16. The length of a rectangle is $(x - 2)$ feet and the width is $(x^2 - 1)$ feet. Which polynomial expression represents the area of the rectangle?

A. $-x^2 - x + 2$

B. $x^3 - 2x^2 - x + 2$

C. $x^3 + 3x + 2$

D. $x^3 - 2x$

17. _____ is the polynomial expression that represents the area of the triangle shown below.

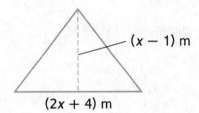

$(x - 1)$ m

$(2x + 4)$ m

18. Justin wants to enclose his rectangular garden space shown below with a fence. Which polynomial expression represents the amount of fencing he needs to enclose the garden?

$(3x + 5)$ ft

$(9x - 1)$ ft

A. $(12x + 4)$ ft

B. $(24x + 8)$ ft

C. $(27x^2 + 42x - 5)$ ft

D. $(27x^2 + 48x - 5)$ ft

This lesson will help you solve problems involving real-world polynomials using various factoring methods. Use it with core lesson 4.2 *Factor Polynomials* to reinforce and apply your knowledge.

Key Concept

People practicing a variety of professions and hobbies write, simplify, and evaluate polynomial expressions. Polynomial expressions can be classified by their number of terms or by the greatest exponential power.

Core Skills & Practice

• Build Lines of Reasoning
• Make Use of Structure

Factoring Out Monomials

To factor a polynomial means to write the polynomial as the product of two or more polynomials.

Directions: Use the trinomial shown below to answer questions 1–2.

$$30x^4y^4 + 45x^2y^3 + 75xy^2$$

1. What is the greatest common factor (GCF) of the trinomial shown?

A. $5xy$

B. $5xy^2$

C. $15xy$

D. $15xy^2$

2. Which of the following is the factored form of the trinomial shown?

A. $5xy(6x^3y^3 + 9xy^2 + 15y)$

B. $5xy^2(6x^3y^2 + 9xy + 15)$

C. $15xy(2x^3y^3 + 3xy^2 + 5y)$

D. $15xy^2(2x^3y^2 + 3xy + 5)$

Directions: Answer the questions below.

3. In the monomial, $11xy^2$, 11 is called the _____.

4. What is the factored form of the binomial $9ab^2 + 18a^2b$?

A. $3(3ab^2 + 6a^2b)$

B. $9(ab^2 + 2a^2b)$

C. $3ab(3b + 6a)$

D. $9ab(b + 2a)$

5. What is the factored form of the expression $16x - 80$?

A. $2(8x - 40)$

B. $4(4x - 20)$

C. $16(x - 5)$

D. $16x(x - 5)$

6. Samantha factored a polynomial so that
$9x^2y - 6y^2 + 12xy = 3xy(3x - 2y + 4)$.

She 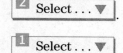 Select . . . ▼ . The correct answer is

Select . . . ▼ .

1 Select . . . ▼   2 Select . . . ▼

A. did not make an error   A. her answer

B. made an error by   B. $3y(3x^2 - 2y + 4x)$
   including x in the GCF

C. made an error by   C. $3x(3x^2 - 2y + 4x)$
   including y in the GCF

D. made an error by   D. $xy(3x^2 - 2y + 4x)$
   including 3 in the GCF

7. A _____ is a polynomial with one term.

8. What is the factored form of the trinomial
$8mn^3 - 24m^2 - 12n^2$?

A. $4(2mn^3 - 6m^2 - 3n^2)$

B. $8(mn^3 - 3m^2 - n^2)$

C. $4mn(2n^2 - 6m - 3n)$

D. $8mn(n^2 - 3m - n)$

9. What is the factored form of the trinomial
$2xy^2 - 5y^2 - 10xy^3$?

A. $2xy(y - 5y - 5y^2)$

B. $2y^2(x - 5 - 5xy)$

C. $y(2xy - 5 - 10xy^2)$

D. $y^2(2x - 5 - 10xy)$

Factoring Quadratic Expressions

Often, real-world situations can be best modeled by quadratic equations. In order to solve quadratic equations, you must first know how to factor quadratic expressions.

Directions: Answer the questions below.

10. A Select . . . ▼ degree polynomial is also
called a quadratic expression.

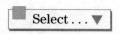 Select . . . ▼

A. first

B. second

C. third

D. fourth

11. The factored form of $x^2 - 5x - 24$
is _____.

12. What is the factored form of $x^2 - 8x + 7$?

A. $(x + 7)(x + 1)$

B. $(x - 7)(x + 1)$

C. $(x - 7)(x - 1)$

D. $(x + 7)(x - 1)$

13. Which of the following is $4x^2 - 4x - 24$ factored completely?

A. $(4x + 8)(x - 3)$

B. $4(x - 3)(x + 2)$

C. $4(x + 2)(x - 3)$

D. $4(x - 3)(x - 2)$

 Test-Taking Tip

Pay special attention to the positive and negative signs of numbers in factors, as this could help you quickly eliminate certain answer choices in a multiple-choice question.

14. Which of the following is $12x^3 + 2x^2 - 10x$ factored completely?

A. $(12x^2 - 10x)(x + 1)$

B. $(2x^2 + 2x)(6x - 5)$

C. $2x(6x + 5)(x - 1)$

D. $2x(6x - 5)(x + 1)$

15. What is the factored form of $6x^2 + 13x - 5$?

A. $(3x - 1)(2x + 5)$

B. $(3x + 1)(2x - 5)$

C. $(3x - 1)(2x - 5)$

D. $(3x + 1)(2x + 5)$

16. Which of the following is $8x^3 + 2x^2 - 3x$ factored completely?

A. $(2x^2 - x)(4x + 3)$

B. $(4x^2 - 3x)(2x - 1)$

C. $x(4x + 3)(2x - 1)$

D. $x(4x - 3)(2x + 1)$

17. A catapult is used to launch water balloons. The height of the balloon after t seconds, in feet, is given by the trinomial $(-16t^2 - 16t + 96)$, in feet. Which of the following shows this trinomial completely factored?

A. $-16(t - 3)(t - 2)$

B. $-16(t + 3)(t - 2)$

C. $-16(t - 3)(t + 2)$

D. $-16(t + 3)(t + 2)$

18. The trinomial shown in the rectangle represents the area of the rectangle in square feet. _____ and _____ are two binomials that could represent the length and width of the rectangle.

$$A = (x^2 + 13x + 42)\,\text{ft}^2$$

This lesson will help you practice solving quadratic equations in different ways. Use it with core lesson 4.3 *Solve Quadratic Equations* to reinforce and apply your knowledge.

Key Concept

Quadratic equations can be solved in several ways. Simple quadratic equations can be solved by inspection. More complex ones can be solved by factoring, completing the square, or using the quadratic formula.

Core Skills & Practices

- Reason Abstractly
- Solve Real-World Problems

Solving a Quadratic Equation by Factoring

Quadratic equations can be used to describe the motion of an object or to calculate areas. Factoring is one way to solve quadratic equations.

Directions: Answer the questions below.

1. What is the positive solution of $x^2 - x - 56 = 0$?

   A. 2

   B. 7

   C. 8

   D. 28

2. Using the zero-product principle, the solution to the equation $(x - 4)(x + 5) = 0$ is _____.

3. The solution of the equation $x^2 - 19x = -90$ is _____.

4. The product of two consecutive integers, x and $x + 1$, is 72. Which quadratic equation can be used to find the positive integer that satisfies the situation?

   A. $(x^2 + 1) + 72 = 0$

   B. $(x^2 + 1) - 72 = 0$

   C. $x^2 + x + 72 = 0$

   D. $x^2 + x - 72 = 0$

5. A landscaping contractor is designing a rectangular paved patio for a customer. In order for the patio to fit in the customer's backyard, the area of the patio must be 288 square feet, and the width of the patio must be 12 feet shorter than its length. What is the width of the patio?

   A. 12 feet

   B. 24 feet

   C. 36 feet

   D. 48 feet

Completing the Square

Completing the square is another way to solve quadratic equations.

Directions: Answer the questions below.

6. What is the value of c such that the equation $x^2 + 14x + c = 0$ is a perfect square trinomial?

A. $c = 7$

B. $c = 14$

C. $c = 28$

D. $c = 49$

7. The equation $x^2 - 8x + 16 = 0$ is a _____ trinomial.

8. The solution that $x^2 - 32x = -256$ and $x^2 - 14x = 32$ have in common is _____.

9. The distance d in feet that a dropped object falls in t seconds is given by the equation $d = 16t^2$. If a ball is dropped from the roof of a building top that is 36 feet tall, how many seconds will it take to reach the ground?

A. 0.5 seconds

B. 1.5 seconds

C. 4 seconds

D. 6 seconds

10. Which equation has no real solution?

A. $x^2 - 10x = -25$

B. $-x^2 = -100$

C. $x^2 - 25 = 0$

D. $x^2 + 49 = 0$

11. In order to solve the quadratic equation $x^2 + 16x + 39 = 0$, first 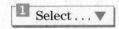. After doing this, you need to 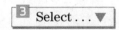 and then 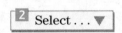. Rewriting the equation as a perfect square gives the equation , which has solutions 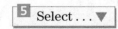.

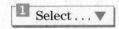 Select . . . ▼

A. add 64 to both sides of the equation

B. divide 16 by 2 to get 8

C. subtract 39 from both sides of the equation

D. square 16 to get 256

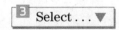 Select . . . ▼

A. add 64 to both sides of the equation

B. divide 16 by 2 to get 8

C. subtract 39 from both sides of the equation

D. square 16 to get 256

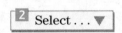 Select . . . ▼

A. add 64 to both sides of the equation

B. divide 16 by 2 to get 8

C. subtract 39 from both sides of the equation

D. square 16 to get 256

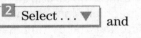 Select . . . ▼

A. $(x + 16)^2 = 64$

B. $(x + 16)^2 = 25$

C. $(x + 8)^2 = 64$

D. $(x + 8)^2 = 25$

Select . . . ▼

A. $-16, 0$

B. $-24, -8$

C. $-13, -3$

D. $-21, -11$

The Quadratic Formula

The quadratic formula allows you to solve any quadratic equation by substituting values into the formula.

Directions: Answer the questions below.

12. How many solutions are there for $3x^2 + 5x - 10 = 0$?

    A. 0

    B. 1

    C. 2

    D. 3

13. The equation $3x^2 + bx = -12$ has no real solutions, so b must be [Select ... ▼] 12.

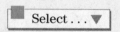 Select ... ▼

    A. greater than

    B. less than

    C. equal to

    D. at least

✓ Test-Taking Tip

When using the discriminant from the quadratic formula, remember that the quadratic equation must first be in the correct form. Once in the correct form, the a, b, and c coefficients can then be used.

14. Which equation has $\dfrac{-5 \pm \sqrt{25 - 4 \times 2 \times (-4)}}{4}$ as its solution?

    A. $-4 + 2x^2 = -5x$

    B. $-4 + 5x^2 = -2x$

    C. $4 + 2x^2 = -5x$

    D. $4 + 5x^2 = -2x$

15. What are the solutions of the equation $3x^2 + 8x - 3 = 0$?

    A. $x = \dfrac{8 \pm 10}{4}$

    B. $x = \dfrac{-8 \pm 10}{4}$

    C. $x = \dfrac{-8 \pm 10}{6}$

    D. $x = \dfrac{-8 \pm \sqrt{28}}{6}$

16. An object is launched upward at a speed of 16 ft/sec from a platform that is 5 ft high. The object's height in feet after t seconds is given by the equation $h = -16t^2 + 16t + 5$. How long will it take before the object lands on the ground?

    A. $t = 0.25$ second

    B. $t = 0.5$ second

    C. $t = 1$ second

    D. $t = 1.25$ seconds

This lesson will help you practice using all four operations to evaluate rational expressions. Use it with core lesson 4.4 *Evaluate Rational Expressions* to reinforce and apply your knowledge.

Key Concept	Core Skills & Practices
A rational expression is a ratio of two polynomials. Rational expressions are similar to fractions and can be simplified, multiplied, divided, added, and subtracted using methods similar to those for fractions.	• Evaluate Expressions • Perform Operations

Simplifying Rational Expressions

Rational expressions are used to solve problems in everyday life, so it's important to know how to simplify them.

Directions: Answer the questions below.

1. Which of the following is a rational expression?

A. $\dfrac{3}{2^x}$

B. $\dfrac{3t^2u^4v}{r^2s^4}$

C. $\dfrac{6 + xyz}{\dfrac{(2xz)^2}{4x^2z^2} - 1}$

D. $\dfrac{2p + 1}{\sqrt{p} - 2}$

2. Evaluate the rational expression $\dfrac{2x^2 + 3x - 2}{-x + 4}$ when $x = -2$.

A. $\dfrac{-8}{3}$

B. 0

C. 6

D. 2

3. Find the restricted values for $\dfrac{2x}{x^3 - 9x}$.

A. $x = -3$ and $x = 3$

B. $x = -3$ and $x = 0$

C. $x = -3$, $x = 0$, and $x = 3$

D. $x = 0$

4. Which shows the rational expression $\dfrac{x^3 - 4x}{x^2 - 4}$ correctly simplified with its restricted values?

A. x; $x \neq -2$, and $x \neq 2$

B. $x - 2$; $x \neq 4$

C. 1; $x \neq -2$

D. $x + 2$; $x \neq 0$

5. If $x = -3$, the value of the rational expression $\dfrac{x + 5}{x^2 - 9}$ is ▢ Select . . . ▼ .

▢ Select . . . ▼

A. $\dfrac{-1}{9}$

B. 2

C. 9

D. undefined

6. The _____ for the rational expression $\dfrac{x^2 + 4x}{x^2 - 16}$ are -4 and 4 because a rational expression is undefined when the denominator is equal to 0.

Multiplying and Dividing Rational Expressions

Multiplying and dividing rational expressions is similar to performing these same operations with fractions.

Directions: Answer the questions below.

7. What should you do first when multiplying rational expressions?

   A. Find the Least Common Denominator.

   B. Rewrite the expression as multiplication by the reciprocal.

   C. Simplify each rational expression.

   D. Add the denominators.

Directions: Use the expression $\dfrac{x^2 + 2x - 8}{2x^2 + 8x} \times \dfrac{x^2 + x - 12}{x^2 - 3x}$ to solve problems 8–10.

8. Find the restricted values for the expression.

   A. $x \neq 0$

   B. $x \neq -4, x \neq 0,$ and $x \neq 3$

   C. $x \neq -4$ and $x \neq 3$

   D. $x \neq 0$ and $x \neq -4$

9. The expression simplified is _____.

10. Evaluate the expression when $x = -3$ and when $x = 3$.

    A. $\dfrac{-5}{18}, \dfrac{7}{18}$

    B. $1\dfrac{5}{18}, \dfrac{7}{18}$

    C. $x \neq -3, \dfrac{-7}{18}$

    D. $\dfrac{-5}{18}, x \neq 3$

11. Find the restricted values for $\dfrac{x^3 + x^2 - 9x - 9}{x^2 + x - 12} \div \dfrac{x^2 + 4x + 3}{x^2 - 4}$.

    A. $x \neq -4, x \neq -2, x \neq 2,$ and $x \neq 3$

    B. $x \neq -4, x \neq -3, x \neq -2, x \neq -1, x \neq 2,$ and $x \neq 3$

    C. $x \neq -3, x \neq -1$

    D. $x \neq -4, x \neq -3, x \neq -2, x \neq -1$

12. The area of a rectangular swimming pool can be expressed by the expression $x^2 + 5x + 6$. Its width is $x + 2$. Therefore, the expression that describes the length of the pool is _____.

13. A rectangular swimming pool has a volume of $18x^3 + 3x^2 - 3x$ and a depth of $2x + 1$. Therefore, the area of the pool in terms of x is _____.

Adding and Subtracting Rational Expressions

Adding and subtracting rational expressions is similar to performing these same operations with fractions.

Directions: Answer the questions below.

14. What is the sum of $\frac{x+3}{x-6} + \frac{x-4}{x+2}$?

A. $\frac{2x-1}{x^2-4x-12}$

B. $\frac{x^2-x-12}{x^2-4x-12}$

C. $\frac{2x^2-5x-26}{x^2-4x-12}$

D. $\frac{2x^2-5x+30}{x^2-4x-12}$

15. Evaluate $\frac{2x+3}{x^2-16} - \frac{4x-4}{x^2-1}$ for $x = 0$ and $x = 4$.

A. $\frac{-67}{16}, \frac{165}{15}$

B. undefined, $\frac{165}{15}$

C. $\frac{-67}{16}$, undefined

D. undefined, undefined

16. John can finish landscaping his yard in x hours. His son can complete the landscaping job in $x + 2$ hours. The expression 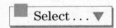 shows how much of the job they can complete in one hour working together.

Select . . . ▼

A. $\frac{2}{x^2+2x}$

B. $\frac{2x+2}{x^2+2x}$

C. $\frac{2}{2x+2}$

D. $\frac{2x+2}{x+2}$

17. It takes Barbara x hours to build a toolshed. Nick can build the same toolshed in $x - 3$ hours. Allyson needs $x + 2$ hours to build the toolshed. The expression 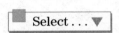 shows how much of the toolshed Allyson and Nick can build in an hour if they are working together.

Select . . . ▼

A. $\frac{2x-3}{x^2-3x}$

B. $\frac{2x+2}{x^2+2x}$

C. $\frac{2x-1}{x^2-x-6}$

D. $\frac{2}{2x-1}$

18. George and Marge use the same looped route to exercise. George takes x hours to rollerblade around the loop at a constant speed. Marge starts where George does but bikes in the opposite direction at a constant speed. They will get $\frac{2x+1}{x^2+x}$ far in one hour. The expression to describe how long it will take Marge to complete the loop is _____.

19. The boat called the Rapids starts at Port Seeyalater and moves down the river with the current towards Port Inawhile. The boat Bayou starts at Port Inawhile up the river against the current towards Seeyalater. The Rapids takes x hours to go from Seeyalater to Inawhile. If the Rapids is twice as fast as the Bayou, the expression representing how close they will be to crossing each other's path in an hour is _____. If it takes the Rapids 2 hours to complete the trip, then it will take them _____ hours to cross each other's path.

 Test-Taking Tip

Using a diagram to model a story problem can show you which objects are moving in which direction.

This lesson will help you practice analyzing slope or rate of change in real-world scenarios. Use it with core lesson 5.1 *Interpret Slope* to reinforce and apply your knowledge.

Key Concept

Slope, a measure of the steepness of a line, is the ratio of vertical change to horizontal change (or rise over run). For lines that represent proportional relationships, the slope of the line is equal to the unit rate.

Core Skills & Practices

• Make Use of Structure
• Use Ratio Reasoning

Points and Lines in the Coordinate Plane

The coordinate plane is a convenient way to plot points from an equation or table.

Directions: Answer the questions below.

1. Determine which quadrant each point would be plotted in. Then write the letter under the appropriate heading.

   A. Negative x-coordinate and positive y-coordinate

   B. $(2, 1)$

   C. $(1, -1)$

   D. Positive x-coordinate and negative y-coordinate

   E. $(-2, -3)$

   | Quadrant I |
   | Quadrant III |
   | Quadrant II |
   | Quadrant IV |

2. Which ordered pair is a solution to the equation of the line pictured below?

   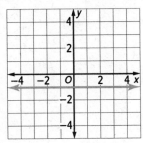

   A. $(-3, 2)$

   B. $(-1, 2)$

   C. $(4, -1)$

   D. $(-2, -3)$

3. Plot three points from the equation of the line $y = -x$.

   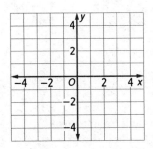

4. Table _____ represents the equation $y = -x + 2$.

A.		B.		C.		D.	
x	y	x	y	x	y	x	y
1	3	0	2	-2	0	-1	3
2	4	1	3	-1	-1	0	2
3	5	2	4	0	-2	1	1

The Slope of a Line

The slope of a line can describe speed, pay rates, interest rates, and the cost of items.

Directions: Answer the questions below.

5. Helen looked at the table below and used the slope formula for two points to calculate the slope. Helen got -2 for the slope. This is wrong. What is the correct calculation of slope?

x	y
−2	−4
−1	−2
0	0
1	2

Helen's calculation:
$$\frac{(-4 + 2)}{(-1 + 2)} = -2$$

A. $\dfrac{(-4 - 2)}{(-1 - 2)} = 2$

B. $\dfrac{(-4 + 2)}{(-2 + 1)} = 2$

C. $\dfrac{(-1 + 2)}{(-4 + 2)} = -\dfrac{1}{2}$

D. $\dfrac{(-2 + 1)}{(-4 + 2)} = \dfrac{1}{2}$

✓ Test-Taking Tip

There are many formulas that you will have to use while taking the test. Don't forget to click the Formula Sheet icon to help you remember the various formulas.

6. Which graph has a slope of 0.75?

A.
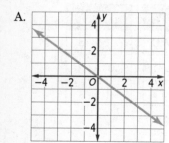

B.
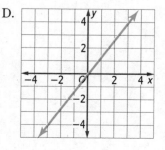

C.
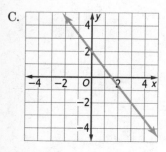

D.

7. The graphs below represent the cost to download music from different Web sites. The number of albums is x and the cost of downloading them is y. Write the names of the Web sites in order of least to most expensive.

Web site A
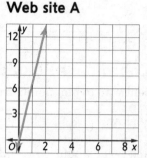

Least Expensive
Most Expensive

Web site B
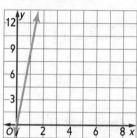

Web site C
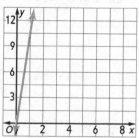

Web site D
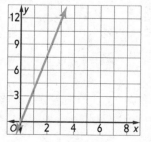

Slope as a Unit Rate

Knowing the slope in a real-world proportional relationship helps you better understand the relationship between the variables.

Directions: Answer the questions below.

8. The cost, y, of hiring two different painters for x hours are represented in the graph and equation below. Select the statement that is false.

Painter A

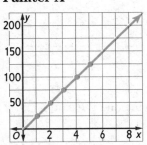

Painter B

$y = 35x$

A. Painter A is less expensive than Painter B.

B. Painter A charges by the hour.

C. Painter B charges $10 more an hour than Painter A.

D. Painter A charges $5 less an hour than Painter B.

9. Two rowers are rowing their boats at constant speeds against the current. For Rower A, the equation $y = 10x$, describes miles traveled in x hours. Rower B's relationship is described in the table below.

Rower B

x hours	y miles
1	15
2	30
3	45

After two and a half hours, rower _____ travels farther by _____ miles.

10. What is the unit rate associated with the table below?

Time in minutes x	Words texted y
3	150
6	300
9	450

A. 3 minutes per 150 words

B. 1 minute per 50 words

C. 50 words per minute

D. 25 words per minute

✓ Test-Taking Tip

For difficult questions that give you choices for the correct answer, eliminate the answers you know are incorrect.

---

11. Which choice describes the unit rate of the graph below?

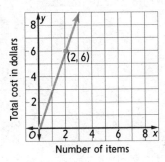

A. $1 per 3 items

B. 6 items per $2

C. 0 items per $0

D. $3 per item

This lesson will help you practice writing the equation of a line from points, slopes, or real-world scenarios. Use it with core lesson 5.2 *Write the Equation of a Line* to reinforce and apply your knowledge.

Key Concept

The equation of a line can be written in many different ways. You can use given information about the line to determine the best way to write the equation.

Core Skills & Practices

- Build Solution Pathways
- Model with Mathematics

Using Slope and y-Intercept

You can use the slope-intercept form to express real life scenarios that have a membership fee and the same cost per item or service.

Directions: Answer the questions below.

1. What is the equation of the line written in standard form? $y = 0.25x - 5$

A. $x - 4y = 20$

B. $0.25x - y = 5$

C. $y + 5 = 0.25x$

D. $-4x + y = 20$

2. Write the equation of a line to express the linear relationship with a unit rate of 3 and an initial value of 2.

A. $3x + y = 2$

B. $y + 2 = 3x$

C. $y = 2x + 3$

D. $y = 3x + 2$

3. Which equation passes through the point (2, 4) and a slope of 3?

A. $y = 3x$

B. $y = 3x - 2$

C. $y - 3 = 4(x - 2)$

D. $y - 2 = 3(x - 4)$

4. Write the equation of a line that best fits the linear scenario of purchasing video rentals for $1 per video and paying $8 total for 3 videos and a one-time membership fee.

A. $y = x$

B. $y = 3x$

C. $y = x + 5$

D. $y = 11$

Test-Taking Tip

Highlight or underline the important information that will help you to solve the problem.

5. Which equation of a line represents the linear relationship of the total cost y of purchasing pounds of food x in bulk for $4 per pound and an initial membership fee of $8?

A. $4x + 8y = 0$

B. $y = 4x + 8$

C. $8x + 4y = 0$

D. $y = 8x + 4$

Using Two Distinct Points

Knowing just two data points for a linear relationship allows you to write the equation of the line.

Directions: Answer the questions below.

6. A line contains the points $(-5, -1)$ and $(-3, -2)$. Write the equation of the line in slope-intercept form.

   A. $y + 2 = -2(x + 3)$

   B. $y = -2x - 2$

   C. $y + 2 = -\frac{1}{2}x(x + 3)$

   D. $y = -\frac{1}{2}x - \frac{7}{2}$

7. Which linear equation is represented in the graph below?

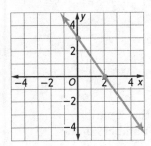

   A. $2x + 3y = 6$

   B. $3x + 2y = 6$

   C. $2x - 3y = 6$

   D. $3x - 2y = 6$

8. What is the equation of the line written in slope-intercept form that contains the points $(-2, 4)$ and $(-1, 2)$?

   A. $y = -2x$

   B. $y = -2x + 8$

   C. $y = 2x$

   D. $y - 2 = 2(x + 1)$

9. When an equation of a line is written in standard form $Ax + By = C$, there are two points you can easily find to calculate the slope. What are the x- and y-intercepts of this line?

   The x-intercept is

   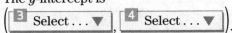,

   The y-intercept is

   .

   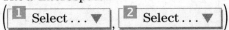

   1 Select . . . ▼   **2** Select . . . ▼

   A. 0   A. 0

   B. $\frac{C}{A}$ B. $\frac{C}{B}$

   C. $\frac{A}{C}$ C. $\frac{B}{C}$

   D. $\frac{C}{B}$ D. $\frac{C}{A}$

   3 Select . . . ▼   **4** Select . . . ▼

   A. 0   A. 0

   B. $\frac{C}{A}$ B. $\frac{C}{B}$

   C. $\frac{A}{C}$ C. $\frac{B}{C}$

   D. $\frac{C}{B}$ D. $\frac{C}{A}$

10. In all three forms, write the equation of the line represented by the graph below. Write the answers on the lines below.

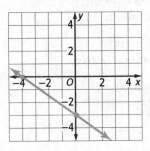

   Standard Form _____$x +$ _____$y =$ _____

   Slope-Intercept $y =$ _____$x +$ _____

   Point-Slope Form $y -$ _____$= m(x -$ _____$)$

Using Tables

Tables are a convenient way to record data to be converted to a graph.

Directions: Answer the questions below.

11. Fill in the table with the points shown on the graph pictured below.

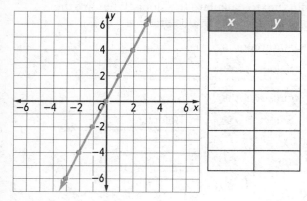

x	y

12. The following data was recorded in the table below for a 100 question exam. Circle the statement that misinterprets the data from the table.

x (the student's hours spent studying at home)	y (the student's correct answers on exam)
0	7
1	16
2	25
3	34
4	43
5	52

A. The student has to study at least 7 hours at home to get 70 correct answers.

B. If the student does not study at home, he will get all the questions incorrect.

C. After three hours of studying at home, the student will get 34 questions correct.

D. The student will get approximately 9 questions correct for every hour of studying at home.

13. Fill in the table of values for the line $4x + 6y = 24$.

x	y
0	
	2

14. _____ is the equation of the line represented by the table.

x	y
1	
2	6
3	9
4	12
5	15
6	

15. Finish the table of values for the equation $y = 3x - 1$. Write the missing numbers in the table.

x	y
−2	
−1	
0	−1
1	2
2	5
3	
4	

16. Complete the table with any 5 points from the graph pictured below. Write the coordinates in the table.

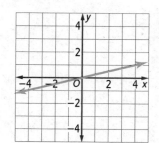

x	y

This lesson will help you learn how to graph equations on a coordinate plane. Use it with core lesson 5.3 *Graph Linear Equations* to reinforce and apply your knowledge.

Key Concept

You can visualize how two variables in an equation are related by graphing the equation. Solutions of a linear equation can be plotted as ordered pairs on the coordinate plane. You can also use the special forms of linear equations to graph them.

Core Skills & Practices

- Solve Linear Equations
- Interpret Graphs

Using Ordered Pairs

You can use ordered pairs to graph points of relationships between two measurable variables, such as time vs. distance, products sold vs. profit, number of items vs. cost of those items, etc.

Directions: Answer the questions below.

1. Which table shows values from the equation $12x - y = 4$?

A. **Table A**

x	y
0	−4
1	−8
2	−20

B. **Table B**

x	y
0	−4
1	8
2	20

C. **Table C**

x	y
0	4
1	8
2	20

D. **Table D**

x	y
0	4
1	−8
2	−20

2. The table below shows ordered pairs for the equation $y = -2x - 3$. Fill in the missing values.

x	y
−2	_____
_____	5
4	_____

3. Choose the graph represented by the ordered pairs from the table below.

x	y
0	1
1	4

A.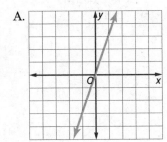

B.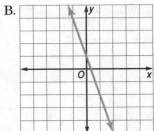

C.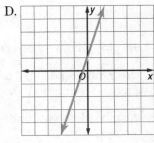

D.

Using Slope-Intercept Form

Linear equations and graphs are used in many different ways to monitor data. Linear equations can predict an outcome based on a constant slope.

Directions: Answer the questions below.

4. Which graph has a negative slope but positive *y*-intercept?

A.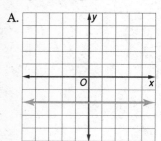

B.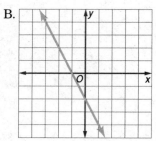

C.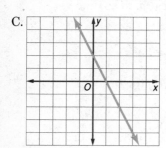

D.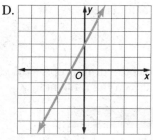

5. Harold graphed the equation $y = -2x + 2$ below. Did he make an error using the slope and *y*-intercept? If so, how?

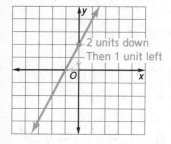

A. Yes, he plotted the incorrect *y*-intercept

B. Yes, after moving down 2 units, he should have moved to the right one unit.

C. Yes, he moved down 2 units from the *y*-intercept.

D. No mistake was made.

6. Gina can type a text message at a rate of 100 words per minute. Her friend can type a text at 90 words per minute. Gina made a linear graph showing the relationship between *x* minutes and the number of words, *y*, she can text. Her friend looked at Gina's graph and realized that the graph representing her situation would have the same [1] Select ... ▼ as Gina's, but a different [2] Select ... ▼ .

[1] Select ... ▼

A. slope

B. *y*-intercept

[2] Select ... ▼

A. slope

B. *y*-intercept

7. Choose the scenario whose graph has a decreasing slope.

A. A company sells an item for $12. Let *x* be the number of items sold and *y* be the total sales.

B. Kate runs 6 miles per hour. She records the number of miles *y* she runs in *x* minutes.

C. Ted bought 100 shirts for $5 each. For 20 consecutive days, he returned 5 shirts each day and got his money back. Let *x* be number of days Ted returned shirts and *y* be the total amount spent on the shirts.

D. Brian is mountain biking up and down a mountain, keeping a constant speed of 10 miles per hour. Let *x* be the time spent mountain biking and *y* be the total distance biked.

8. Cecilia said that the scenario of buying socks that cost $4.00 each and are on sale for Buy One Get One Free could represent the table of data below. Layla said that it could represent the data only if the socks were not on sale. After further discussion, they both agreed that

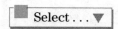 Select . . . ▼ is correct.

x	y
0	0
1	4
2	8

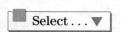 Select . . . ▼

A. Cecilia

B. Layla

9. Sarah is jumping rope at 100 jumps per minute. If she changes to 80 jumps per minute, the slope of her rate will ☐ Select . . . ▼ .

☐ Select . . . ▼

A. become more steep

B. become less steep

C. have no change

10. According to the graph pictured below, Matt joined a music sharing program that costs $_____ initially and $_____ for every song he downloads.

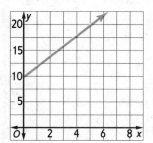

11. Tom is swimming laps to raise money for a charity. He donated $60 and earns $1 for each lap that he swims. Plot 5 points onto the coordinate plane below, starting with the *y*-intercept, which will represent how much money he will raise based on how many laps he swims. Your points should be plotted at intersecting grid lines.

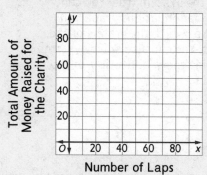

✓ **Test-Taking Tip**

When graphing, make sure that you are aware of the scale on each axis so that the points you draw are accurate to that scale.

12. Which scenario does not fit the following graph?

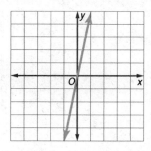

A. running 5 miles per hour

B. taking $5 out of a savings account every week

C. number of items sold at a $5 store

D. filling bags with 5 candy bars

This lesson will help you practice using the graphing, substitution, and elimination methods to find the solution to a system of linear equations. Use it with core lesson 5.4 *Solve Systems of Linear Equations* to reinforce and apply your knowledge.

Key Concept

Just as a solution of an equation is a value that makes the equation true, a solution of a system of equations is a set of values that makes all of the equations in the system true. You can solve systems of linear equations graphically by finding the point at which the graphs of the equations intersect. You can also solve systems algebraically, by using the substitution, or the elimination method.

Core Skills & Practices

- Represent Real-World Problems
- Solve Pairs of Linear Equations

The Graphing Method

Solving a system of equations by graphing is a visual way to determine the solution to the problem.

Directions: Use the graphing method to answer the questions below.

1. Which graph represents the solution to the system of linear equations below?

$$x + 2y = 18$$

$$-x + y = 12$$

A.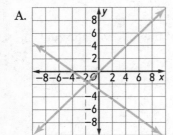

C.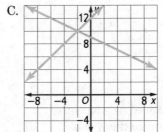

B.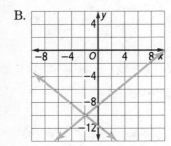

D.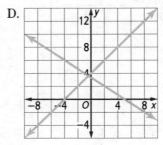

2. Tow-A-Way charges a $25 fee plus $1.25 per mile to tow a car. Haul-Ur-Car charges $1.50 per mile plus a $15 fee to tow a car. Using the graphing method to solve, you would graph the equation _____ for Tow-A-Way and the equation _____ for Haul-Ur-Car, and then determine the solution from the graph.

The Substitution Method

Solving a system of equations by the substitution method is one algebraic way to solve this type of problem.

Directions: Use the substitution method to answer the questions below.

3. The system of linear equations below has

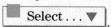 Select . . . ▼ .

$4x + 2y = 5$

$2x + y = 3$

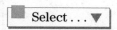 Select . . . ▼

   A. one solution

   B. no solution

   C. infinitely many solutions

   D. two solutions

4. Which of the following is the solution to the system below?

   $x - 4y = 1$

   $3x - y = 3$

   A. $(0, -1)$

   B. $(-1, 0)$

   C. $(0, 1)$

   D. $(1, 0)$

5. Diana and Megan are saving money in bank accounts. Diana starts with $100 and saves $4.50 per week, while Megan starts with $20 and saves $12.50 per week. After how many weeks will the two bank balances be equal? How much money will be in each account?

   A. (8 weeks, $136)

   B. (10 weeks, $145)

   C. (12 weeks, $154)

   D. (14 weeks, $195)

6. Play World charges $25.95 for admission and $1.15 per ride. Fun Land charges $19.95 for admission and $1.25 per ride. For how many rides is the price the same, and what is that price?

   A. (30 rides, $60.45)

   B. (40 rides, $71.95)

   C. (50 rides, $82.45)

   D. (60 rides, $94.95)

7. Water for You sells 12,500 containers of water per month. Drink Up sells 8,000 containers of water per month. After an advertising campaign, Drink Up sees an increase of 40 containers of water sales per month while Water for You has a decrease of 50 containers of water sales per month. Assuming that the trend of increasing and decreasing continues, the equation _____ would represent Water for You and the equation _____ would represent Drink Up. After solving this system of equations, it is found that it will take _____ months for each company to sell _____ containers a month.

 Test-Taking Tip

When trying to solve a problem that involves a system of linear equations, first determine what you are trying to solve for, and use that information to define the variables. Then, write the equations using those variables to model the data in the problem. Last, solve the system.

The Elimination Method

The elimination method is another algebraic method used to solve a system of equations.

Directions: Use the elimination method to answer the questions below.

8. Solve the system below by elimination. This system has 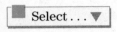 Select ... ▼ .

    $4x - 4y = 12$

    $7x - 7y = 21$

    [ ] Select ... ▼

    A. one solution

    B. no solution

    C. infinitely many solutions

    D. two solutions

9. Which of the following is the solution to the system below?

    $3x + 5y = -64$

    $4x + 3y = -56$

    A. $(-8, -8)$

    B. $(-8, 8)$

    C. $(8, -8)$

    D. $(8, 8)$

10. Tickets to a school play cost $5 per student and $8 per adult. How many of each kind of ticket were sold if 500 tickets were sold, and the play brought in a total of $3,700 in ticket sales?

    A. 100 student tickets, 400 adult tickets

    B. 200 student tickets, 300 adult tickets

    C. 300 student tickets, 200 adult tickets

    D. 400 student tickets, 100 adult tickets

11. Robert buys 5 pairs of jeans and 9 long-sleeve T-shirts. Juan buys 2 pairs of jeans and 6 long-sleeve T-shirts. Juan spends $208, and Robert spends $376. How much does each pair of jeans and each long-sleeve T-shirt cost?

    A. $24 for jeans, $32 for a T-shirt

    B. $28 for jeans, $38 for a T-shirt

    C. $32 for jeans, $24 for a T-shirt

    D. $38 for jeans, $28 for a T-shirt

12. On a 1,200-mile trip, a plane trip took 5 hours flying in the direction of the wind. On the return trip flying against the wind, the trip took 6 hours. Use x as the plane speed and y as the wind speed to determine which is the solution to the system of equations.

    A. (20 mph, 220 mph)

    B. (220 mph, 20 mph)

    C. (-20 mph, 220 mph)

    D. (220 mph, -20 mph)

This lesson will introduce you to functions and how to recognize functions of many types in graphs, tables, or algebraic equations. Use it with core lesson 6.1 *Identify a Function* to reinforce and apply your knowledge.

Key Concept

A function assigns exactly one output for each input. The inputs of a function are a given set, and the outputs for this function create another set. The outputs are the result of operations performed on the set of inputs. A good way to identify a function is to use the Vertical Line Test.

Core Skills & Practices

- Use Math Tools Appropriately
- Solve Real-World Problems

Functions

Functions can represent various relationships between inputs and outputs.

Directions: Answer the questions below.

1. Which of the graphs below represent one-to-one functions?

A.

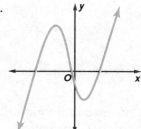

C.

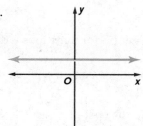

B.

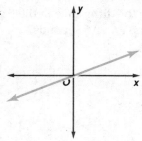

D.

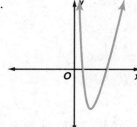

Directions: Use the following scenario for questions 2 and 3.

The function $f(x) = 0.49x + 44.95$ describes the total cost for a company to print flyers, where x is the number of flyers printed.

2. What is the cost to print 1 flyer?

   A. $0.49

   B. $4.90

   C. $44.95

   D. $45.44

3. _____ is the total cost of printing 250 flyers.

Linear and Quadratic Functions

A linear function would describe a set price for a number of products sold.
A quadratic function is the height of an object thrown in the air.

Directions: Answer the questions below.

4. Using the functions $f(x) = x^2 - 5x + 6$ and $g(x) = -\frac{1}{2}x + 3$, order the following values from least to greatest.

$g(-3)$

$f(0)$

$g(0)$

$f(3)$

Least
Greatest

5. Which of the following is true for the function $f(x) = 3x^2 - 4x + 1$?

A. $f(0) > f(1)$

B. $f(0) > f(-1)$

C. $f(1) > f(-1)$

D. $f(1) > f(0)$

Directions: Use the following scenario for questions 6 through 7.

The cost of buying a gallon of gasoline can be represented by the function $f(x) = 3.18x$ where x is the number of gallons of gasoline purchased.

6. Evaluate the function to find the cost of 5 gallons and the cost of 12 gallons of gasoline.

A. 5 gallons: $19.00; 12 gallons: $38.16

B. 5 gallons: $15.90; 12 gallons: $45.60

C. 5 gallons: $19.00; 12 gallons: $45.60

D. 5 gallons: $15.90; 12 gallons: $38.16

7. Which of the following describes the properties of the function?

A. linear, not one-to-one

B. linear, one-to-one

C. quadratic, not one-to-one

D. quadratic, one-to-one

8. The height of a ball thrown upward into the air from a 98-meter-tall building can be represented by the function, $f(t) = -4.9t^2 + 19.6t + 98$ where t is time in seconds. What is the difference in the height of the ball between 1.5 and 2.5 seconds?

A. 0 meters

B. 1 meter

C. 5 meters

D. 10 meters

Functions in the Coordinate Plane

Graphing a function is a way to visually express patterns and behaviors between inputs and outputs.

9. Determine the values for the piecewise function below when $x = -3$, $x = -1$, $x = 0$, and $x = 3$. Write the answers in the table.

$$f(x) = \begin{cases} 5x - 3 \text{ when } x < -1 \\ 8 \text{ when } x = -1 \\ -\frac{2}{3}x + 4 \text{ when } x > -1 \end{cases}$$

x	f(x)
−3	
−1	
0	
3	

C.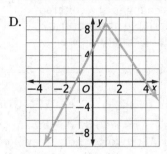

D.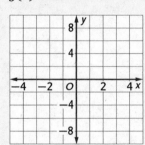

10. Plot the points for the function $f(x) = 2x^2 - 3x - 4$ when $x = -2$, $x = 0$, and $x = 2$.

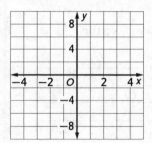

11. Which graph represents the piecewise function?

$$h(x) = \begin{cases} 4x + 5, \ x \leq 1 \\ -4x + 5, \ x > 1 \end{cases}$$

A.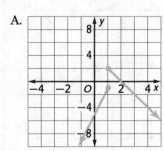

B.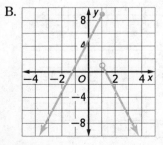

12. Plot 3 points on the line for the linear function:
$g(x) = 2x - 5$

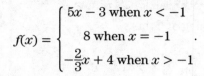

✓ Test-Taking Tip

When constructing a graph, make sure to label the axes clearly. When reading a graph, look carefully at the axes to make sure you are interpreting the scale correctly.

This lesson will help you distinguish between linear and quadratic functions by analyzing common differences, or identifying graphs. Use it with core lesson 6.2 *Identify Linear and Quadratic Functions* to reinforce and apply your knowledge.

Key Concept

Linear and quadratic functions express a relationship between two variables — one independent and the other dependent. As the independent variable changes, the dependent variable of linear functions changes at a constant rate while the dependent variable of quadratic functions does not change at a constant rate.

Core Skills & Practices

- Critique the Reasoning of Others

Evaluating Linear and Quadratic Functions

Linear and quadratic functions are used to model data in health, economics, nature, and just about everywhere else. So, it is important to know how to evaluate them for information.

Directions: Answer the questions below.

1. What is the common consecutive difference for the function $f(x) = \frac{1}{4}x + 4$?

A. $\frac{1}{4}$

B. 4

C. $-\frac{1}{4}$

D. -4

2. The common consecutive difference for the function $f(x) = 2x^2 + 1$ is [1] Select . . . ▼

The common consecutive difference was found on the [2] Select . . . ▼ set of consecutive differences, therefore the function is

[3] Select . . . ▼

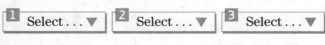

[1] Select . . . ▼	[2] Select . . . ▼	[3] Select . . . ▼
A. 1	A. 1st	A. constant
B. 2	B. 2nd	B. linear
C. 4	C. 3rd	C. quadratic
D. 6	D. 4th	D. polynomial of degree 3

3. Which table of values corresponds to the function $f(x) = 11x - 3$?

A.
x	-2	-1	0	1	2	3
$f(x)$	19	16	13	10	7	4

B.
x	-2	-1	0	1	2	3
$f(x)$	-25	-22	-19	-16	-13	-10

C.
x	-2	-1	0	1	2	3
$f(x)$	-25	-14	-3	8	19	30

D.
x	-2	-1	0	1	2	3
$f(x)$	19	8	-3	-14	-25	-36

4. The function $f(x) = -x^2 + 3$ is represented by which of the tables below?

A.
x	-2	-1	0	1	2	3
$f(x)$	-1	2	3	2	-1	-6

B.
x	-2	-1	0	1	2	3
$f(x)$	1	-2	-3	-2	1	6

C.
x	-2	-1	0	1	2	3
$f(x)$	7	4	3	4	7	12

D.
x	-2	-1	0	1	2	3
$f(x)$	-7	-4	-3	-4	-7	-12

Directions: Circle the correct graph.

5. Which graph represents the function $f(x) = x^2 + 2x - 8$?

A.

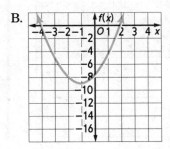

B.

C.

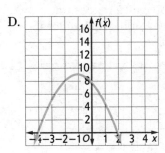

D.

Recognizing Linear and Quadratic Functions

Linear and quadratic functions behave differently. Learning to recognize the differences in their behavior helps you understand the information they are representing.

Directions: Answer the questions below.

6. Find the common consecutive differences of the tables of values to identify whether the table represents a linear function, quadratic function, or neither. Write your answer on the line.

x	f(x)	1st Cons. Diff.	2nd Cons. Diff.
−2	−19		
−1	−9		
0	1		
1	11		
2	21		
3	31		

The table represents a _____.

7. Use common consecutive differences to show $f(x) = -2$ is a linear function. The common difference is _____.

x	f(x)	1st Cons. Diff.
−2		
−1		
0		
1		
2		
3		

✓ **Test-Taking Tip**

Use scrap paper to help you keep your thoughts organized while taking the test. When filling out tables for consecutive differences, subtract the previous y-value from the current y-value. Then place that difference in the row that corresponds to the previous y-value.

8. What type of function has a common set of 2nd consecutive differences?

A. constant

B. linear

C. quadratic

D. polynomial of degree 3

9. Find the common consecutive differences for each of the functions below. Put the differences in order from least to greatest.

x	f(x)
−3	−2
−2	2
−1	6
0	10
1	14
2	18

x	g(x)
−3	1
−2	3
−1	5
0	7
1	9
2	11

x	h(x)
−3	5
−2	5
−1	5
0	5
1	5
2	5

 f(x)

 g(x)

h(x)

Least

Greatest

10. The following table represents a polynomial function. Complete the table to determine what kind of function it is.

A. linear

B. quadratic

C. polynomial of degree 3

D. polynomial of degree 4

x	f(x)	1st Cons. Diff.	2nd Cons. Diff.	3rd Cons. Diff.	4th Cons. Diff.
−3	83				
−2	18				
−1	3				
0	2				
1	3				
2	18				
3	83				

11. Helen says that the tables below represent data from the same quadratic function. Use common consecutive differences to determine whether Helen is right or wrong.

Table A	
x	f(x)
−2	4
−1	1
0	0
1	1
2	4
3	9

Table B	
x	f(x)
4	30
5	50
6	75
7	100
8	150
9	200

Helen is _____.

This lesson will help you understand the key features of a graph and what the features mean in a real-world situation. Use it with core lesson 6.3 *Identify Key Features of a Graph* to reinforce and apply your knowledge.

Key Concept

Key features that can be identified from graphs include intercepts, positive and negative intervals, increasing and decreasing intervals, relative minimums and maximums, end behavior, symmetry, and periodicity. You can sketch graphs if you know or can determine the key features to represent the visual features of the graph.

Core Skills & Practices

- Make Use of Structure
- Gather Information

Key Features

Graphs are used in many fields. It's important to know how to read and interpret their key features.

Directions: Answer the questions below.

1. Determine the values of the following key features of the graph below. Write your answers in the boxes.

none	16
15	−3, 5

$x = 1$

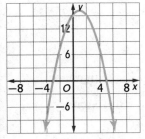

A. line of symmetry [ ]

B. relative minimum [ ]

C. relative maximum [ ]

D. x-intercept(s) [ ]

E. y-intercept(s) [ ]

2. Which of the following is a linear function with x-intercept of −4 and y-intercept of 5? Circle the correct graph.

A.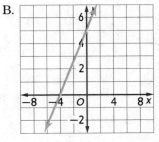

B.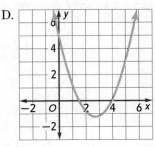

C.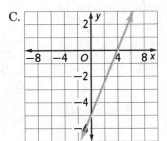

D.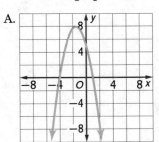

Directions: Use the following graph to answer problems 3 and 4.

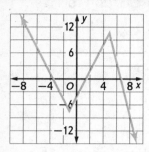

3. Determine the increasing and decreasing intervals for the function.

   A. increasing: all values for x

   B. increasing: $x < 5$, decreasing $x > 5$

   C. increasing: $-1 < x < 5$, decreasing: $x < -1$ and $x > 5$

   D. increasing: $x < -1$ and $x > 5$, decreasing: $-1 < x < 5$

4. Describe the end behavior of the function.

   A. The left end is increasing indefinitely and the right end is decreasing indefinitely.

   B. The left end is increasing indefinitely and the right end is increasing indefinitely.

   C. The left end is decreasing indefinitely and the right end is decreasing indefinitely.

   D. The left end is decreasing indefinitely and the right end is increasing indefinitely.

Directions: Use the following graph to answer problems 5 and 6.

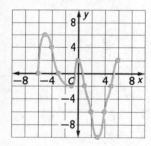

5. The largest relative maximum in the graph is:

   A. 2

   B. 4

   C. 6

   D. 8

6. The smallest relative minimum in the graph is:

   A. −2

   B. −6

   C. −10

   D. −14

7. Identify the key features of the following graph. Write your answers on the lines.

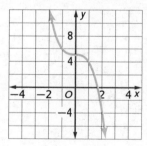

   A. The graph shows _____ symmetry.

   B. The end behavior on the left side extends indefinitely _____.

   C. The end behavior on the right side extends indefinitely _____.

   D. The decreasing interval is _____.

Use Key Features to Draw a Graph

Being able to create graphs can help you visually communicate data and information.

Directions: Answer the questions below.

8. On the following graph, place a point at the x-intercept(s).

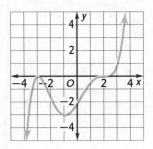

9. On the following graph, place a point at the relative maximum(s) and relative minimum(s).

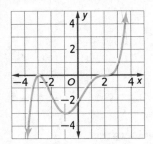

10. What is the end behavior of the following graph?

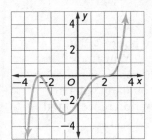

A. extends indefinitely up both to the left and right

B. extends indefinitely down both to the left and right

C. extends indefinitely down to the left and up indefinitely to the right

D. extends indefinitely up to the left and down indefinitely to the right

11. Determine which graph of a function has relative maximums at y-values 0 and −1, a relative minimum at y-value of −3, an x-intercept at 2, and extends down indefinitely in both directions. Circle the correct graph.

A.
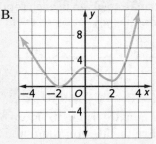

B.
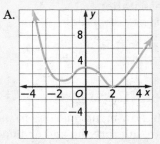

C.
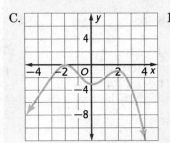

D.
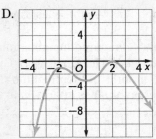

✓ Test-Taking Tip

When looking at the end behavior of a graph, it helps to plot points for any intercepts, draw arrows for end behavior, and then connect the points, paying attention to any maximums and minimums.

This lesson will help you practice comparing functions in different ways. Use it with core lesson 6.4 *Compare Functions* to reinforce and apply your knowledge.

Key Concept
Functions can be represented in many ways – graphs, tables, equations, or verbal descriptions. To compare two or more functions represented in different ways, you will have to use the information given in each representation to determine key features that can be compared.

Core Skills & Practices
• Use Ratio and Rate Reasoning • Make Sense of Problems

Compare Proportional Relationships

Whether you are trying to figure out how many quarters to put into a parking meter or how much you need to earn to live the life you want, you are interpreting and evaluating proportions.

Directions: Answer the questions below.

1. Haylee compared membership fees for two local gyms. The Move More Gym charges $30 for 3 months of membership. The Get Fit Gym membership fees are given in the table below.

Months of Membership	Cost
1	$15
2	$30
3	$45

Which of the following statements is false?

A. The Move More gym costs $10 a month.

B. The Move More Gym costs more per month than the Get Fit Gym.

C. The Move More Gym charges $60 for 6 months.

D. The Get Fit Gym costs $45 for 3 months.

2. Two publishing companies are interested in publishing Kimberly's book. Sell Your Book company will pay Kimberly 20 cents for every book sold. The E-Your Book company will pay Kimberly according to the function represented by the graph below.

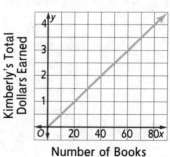

Number of Books Downloaded

Choose the statement that has incorrect information.

A. Kimberly will earn $3 from E-Your Book if 60 books are downloaded.

B. Sell Your Book will pay Kimberly more per book sold than E-Your Book will pay per book downloaded.

C. Kimberly will earn the same amount from both companies when 100 books are sold or downloaded.

D. Kimberly will earn $12 from Sell Your Book if 60 books are sold.

Compare Linear Functions

Linear functions are good models to use to express cost situations where there are a one-time fee and a charge per item.

Directions: Answer the questions below.

3. To reduce expenses in her store, Karen is considering switching to new energy-efficient light bulbs. The graph below shows the purchase price and monthly operating costs for the bulb 13WSR. The purchase price for the 58WTR is $12 and $1.00 per month operating cost.

    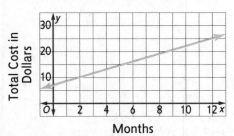

    Which statement is false?

    A. The total cost of bulb 13WSR for 3 months is $11.50.

    B. The 58WTR bulb is more expensive than the 13WSR at 10 months.

    C. After a year, the total cost of the 13WSR is more than the 58WTR.

    D. For the first five months, the 58WTR is more expensive than the 13WSR.

4. Two grocery stores sell almonds by the pound. The Fresh Day grocery store sells almonds for $3 a pound plus a container fee of $2. The Friendly Earth grocery store sells almonds according to the prices shown in the table below.

Pounds of Almonds	Total Cost at Friendly Earth
1	$5
2	$7
3	$9
4	$11

    According to the data for Fresh Day and Friendly Earth, Fresh Day will become more expensive than Friendly Earth when greater than _____ pound(s) of almonds is sold.

5. Shandell is paying an electrician a $30 home-service fee plus $40 per hour for time actually worked. She is also paying a plumber according to the table below.

Plumber Hours Worked	Total Cost
1	55
2	110
3	165
4	220

    Choose the statement that is not true.

    A. For 1 hour of work, the plumber charges less than the electrician.

    B. For 2 hours of work, the plumber and the electrician cost the same amount.

    C. For 3 hours of work, the electrician is more expensive than the plumber.

    D. For 4 hours of work, the electrician is cheaper than the plumber.

     Test-Taking Tip

    If you get confused when comparing a table with a written function rule, try making a table for the function rule. If you're comparing a graph to a table or equation, you can try sketching the data from the table or equation to check your work.

Compare Quadratic Functions

Quadratic functions are used to model scenarios such as the height of an object in motion with respect to time or the height of an object in motion with respect to its horizontal distance.

Directions: Use the information below for questions 6 and 7.

Mr. Bott hits his golf ball off an elevated tee 64 feet from the ground. The height of his golf ball after t seconds is $H(t) = -16t^2 + 120t + 64$. Mrs. Bott hits her golf ball off an elevated tee and the height of her golf ball after t seconds is represented by the graph below.

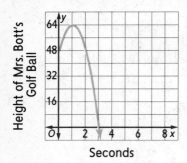

6. Mr. Bott's golf ball will hit the ground after _____ seconds and Mrs. Bott's golf ball will hit the ground after _____ seconds.

7. Mrs. Bott's golf ball will reach a maximum height of _____ feet.

Directions: Use the information below for questions 8 and 9.

The potential weekly profits of an older book and a newer book are shown in the graph and table below, where x is the number of weeks the book is on sale and y is the total profit for the week, in thousands of dollars.

Number of Weeks on Sale (x)	Newer Book's Weekly Total Profit (y)
9	0
11	18
13	28
14	30
14.5	30.25
15	30
16	28
18	18
20	0

Number of Weeks on Sale

8. The [Select . . . ▼] book has the potential to make the most profit in a single week.

   [Select . . . ▼]

   A. newer

   B. older

9. The older book will start making a profit after _____ weeks and stop making a profit after _____ weeks while the newer book will start making a profit after _____ weeks and stop after _____ weeks.

This lesson will help you with calculating perimeter and area of 2-dimensional shapes. Use it with core lesson 7.1 *Compute Perimeter and Area of Polygons* to reinforce and apply your knowledge.

Key Concept	Core Skills & Practices
Formulas can be used to find the perimeter and area of polygons.	• Calculate Area • Perform Operations

Rectangles

Rectangles are a type of polygon with four right angles.

Directions: Answer the questions below.

1. Your driveway is 35 ft long and 24 ft wide. One tub of blacktop sealer costs $18.50 per tub and it covers 450 square feet. How much will it cost to seal your driveway with two coats of sealer?

   A. $18.50

   B. $37.00

   C. $55.50

   D. $74.00

2. The 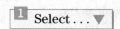 of a rectangle equals 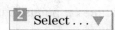 the width plus twice the length.

① Select . . . ▼	② Select . . . ▼
A. area	A. one third
B. perimeter	B. three times
C. height	C. twice
D. diagonal	D. one half

3. _____ is the area of the following rectangle.

   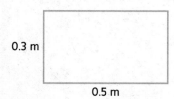

   0.3 m

   0.5 m

4. Eastbridge High School is 80 ft long and 52 ft wide. How many times must a runner run around the school to run a mile? (1 mile = 5,280 ft)

   A. 25 times

   B. 10 times

   C. 20 times

   D. 40 times

5. A box measures 10 in. in height, 18 in. across the front, and 12 in. front to back. Ribbon is being wrapped up the front face, across the top, down the back face, and bottom. How much ribbon is needed to complete one time around the box?

   A. 32 m

   B. 38 m

   C. 40 m

   D. 44 m

6. A cutting board is 30 cm wide. How long must it be to have an area of 1,140 cm²?

   A. 28 cm

   B. 38 cm

   C. 42 cm

   D. 540 cm

Triangles

Triangles are the building blocks of many weight-bearing structures because they are inherently strong and rigid.

Directions: Answer the questions below.

7. Find the perimeter of the triangle.

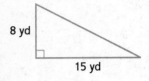

8 yd

15 yd

A. 23 yd

B. 38 yd

C. 40 yd

D. 60 yd

8. The base of a triangular sail will be 3 m wide. How tall must the sail be to have an area of 7.5 m²?

A. 2.5 m

B. 4.5 m

C. 5 m

D. 10.5 m

9. The area of the following triangle is _____.

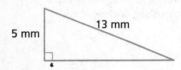

13 mm

5 mm

10. An isosceles triangle has one side 8 in. long and another side that is 3 in. long. What is the perimeter? Hint: Try to draw all possible solutions.

A. 11 in.

B. 14 in.

C. 19 in.

D. 24 in.

11. Write the information for the triangle below.

20 ft   60 ft   150 ft²

Length of side b =

Perimeter =

Area =

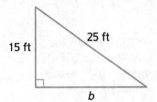

25 ft

15 ft

b

12. The perimeter of the triangle is 27 units. _____ is the length of side a.

7   9

a

13. A helicopter flying at 120 mph takes off at noon and flies south for 180 mi and then west 240 mi before returning, as shown. Fill in the expected arrival times at points A and B as well as the return time.

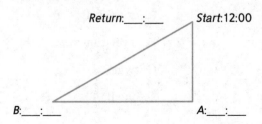

*Return:*___:___   *Start:*12:00

*B:*___:___   *A:*___:___

Parallelograms and Trapezoids

A parallelogram has two sets of opposite parallel sides while a trapezoid has one set of opposite parallel sides.

Directions: Answer the questions below.

14. What is the area of the parallelogram crosswalk?

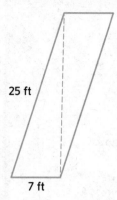

25 ft

7 ft

A. 84 ft²

B. 168 ft²

C. 175 ft²

D. 300 ft²

15. Two 9-in. poles and two 15-in. poles are attached with flexible joints to make a rectangular frame. How much more area does the frame contain when it is in Position A than in Position B?

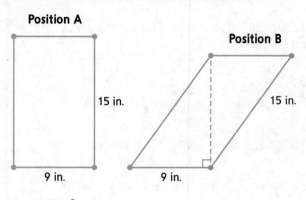

Position A

15 in.

9 in.

Position B

15 in.

9 in.

A. 13.5 in.²

B. 27 in.²

C. 108 in.²

D. 135 in.²

16. You are laying out a flowerbed in the shape of a trapezoid. One base will be 3 m and the height will be 5 m. What must be the length of the other base so that the flowerbed will have an area of 35 m²?

A. 4 m

B. 7 m

C. 9 m

D. 11 m

✔ Test-Taking Tip

If you are asked to find a missing side length of a polygon based on its area or perimeter, first write out the formula for the area or perimeter of that polygon. Then, substitute the information you have and solve for the missing side length.

---

17. What is the length of the line segment labeled h in the parallelogram shown?

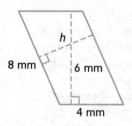

h

8 mm

6 mm

4 mm

A. 3 mm

B. 5.67 mm

C. 6 mm

D. 12 mm

18. What is the height of the trapezoid? _____

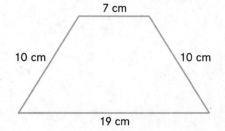

7 cm

10 cm

10 cm

19 cm

This lesson will help you practice using formulas to find circumference and area of circles. Use it with core lesson 7.2 *Compute Circumference and Area of Circles* to reinforce and apply your knowledge.

Key Concept	Core Skills & Practices
You can use formulas to find the circumference and area of circles.	• Perform Operations • Calculate Area

Circumference

The circumference of a circle is the distance around the edge of the circle, similar to the perimeter of a polygon.

Directions: Answer the questions below.

1. A circular racetrack is 600 feet in diameter. To the nearest tenth of a second, how long will it take a car traveling at 150 ft/sec to make one lap around the track? Use 3.14 for π.

   A. 4 sec

   B. 6.3 sec

   C. 12.6 sec

   D. 25.1 sec

2. A circular park is shown in the figure. One path leads around the border of the park, and another path leads straight through the center. The path through the center from the entrance to the fountain is 800 m long. To the nearest tenth of a meter, how much longer is the path from the entrance to the fountain by going around the border of the park? Use 3.14 for π.

   Entrance

   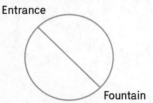

   Fountain

   A. 228 m

   B. 456 m

   C. 1256 m

   D. 1712 m

3. A bicycle wheel has a diameter of 0.67 m. To the nearest revolution, how many revolutions must the wheels make to cover 1 km? (1 km = 1000 m) Use 3.14 for π.

   A. 238

   B. 318

   C. 475

   D. 670

4. The diagram below shows a large gear and a small gear.

   5 in.   3 in.

   When the large gear turns 15 times, the small gear turns _____ times.

   A. $1\frac{2}{3}$

   B. 25

   C. $41\frac{2}{3}$

   D. 9

5. A stone is dropped into a pool. The water ripples outward at a rate of 2 ft/sec. To the nearest foot, what will be the circumference of the ripple after 15 seconds? Use 3.14 for π.

A. 47 ft

B. 94 ft

C. 141 ft

D. 188 ft

6. The blades of a helicopter are 9 ft long. If the blades spin at the rate of 300 revolutions per minute, the tips of the blades will travel _____ feet in 30 seconds (to the nearest 10 feet), using 3.14 for π.

Area

The area of a circle is the two-dimensional space inside its circumference.

Directions: Answer the questions below.

7. The number π is the ratio of the area of a circle to the square of its 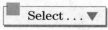 Select ... ▼ .

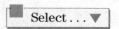 Select ... ▼

A. diameter

B. circumference

C. radius

D. area

8. What is the area of a circle whose circumference is 22π?

A. 11π

B. 121π

C. 363π

D. 484π

9. A chef has been preparing sushi on a circular table with a diameter of 15 in. She decides she needs a table with twice the area of her present table. To the nearest tenth of an inch, the minimum diameter her new table must have is _____ inches, using 3.14 for π.

10. Gregor is covering the top of a circular box with glitter as a stage prop for a play. The box has a radius of 10 in. Each packet of glitter covers an area of 12 in². Therefore, Gregor will need 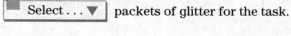 Select ... ▼ packets of glitter for the task.

Select ... ▼

A. 6

B. 27

C. 63

D. 314

11. What is the area of the heart-shaped figure below to the nearest tenth of a square centimeter? Use 3.14 for π.

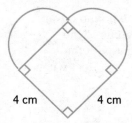

4 cm   4 cm

A. 22.3 cm²

B. 28.6 cm²

C. 41.1 cm²

D. 66.3 cm²

12. What is the area of the shaded part of the figure to the nearest square meter? Use 3.14 for π.

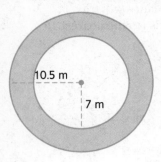

A. 39 m²

B. 192 m²

C. 297 m²

D. 346 m²

13. The area of a circle that has a circumference of 18π is _____ π.

14. Stefan said that the area of a circle with a diameter of 10 is 100π but Raffi said that the area is 25π. Ann worked through the problem with them and they all agreed

that [Select . . . ▼] is correct.

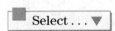

A. Stefan

B. Raffi

C. Ann

D. nobody

Find Radius or Diameter

The formulas for circumference and area can also be used to find a radius or a diameter.

Directions: Answer the questions below.

15. After running 4 laps around her high school's circular track, Sonya had completed a mile. To the nearest foot, what is the radius of the track? Use 3.14 for π (1 mi = 5,280 ft)

A. 210 ft

B. 220 ft

C. 420 ft

D. 840 ft

16. You have 2 lbs of grass seed. Each pound covers 400 ft². To the nearest foot, the radius of the largest circular lawn you can plant with the grass seed that you have is _____ feet when you use 3.14 for π.

17. A Ferris wheel has a circumference of 126 ft. To the nearest foot, what is the highest it lifts you above the ground? Use 3.14 for π.

A. 40 ft

B. 42 ft

C. 80 ft

D. 402 ft

18. Janna used 2 cups of flour to make a pizza with a diameter of 8 in. If she uses $4\frac{1}{2}$ cups of flour, what is the diameter of the pizza she can make?

A. 10.5 in.

B. 12 in.

C. 12.5 in.

D. 18 in.

✓ Test-Taking Tips

It is easy to confuse the radius and the diameter of a circle when making calculations, since both are used to solve problems. The diameter or the radius can be used to find the circumference. The radius is needed to find the area. Note whether you are given the radius or the diameter and, if necessary, convert from one to the other by multiplying or dividing by 2 before making further calculations.

This lesson will help you calculate surface area and volume of three-dimensional objects. Use it with core lesson 7.3 *Compute Surface Area and Volume* to reinforce and apply your knowledge.

Key Concept

The volume of a three-dimensional object is the number of cubic units it takes to fill the object. The surface area of a three-dimensional object is the number of square units it takes to cover all sides of the object.

Core Skills & Practices

• Calculate Volume
• Calculate Surface Area

Rectangular Prisms

Many boxes are shaped like rectangular prisms, so it's useful to know how to find surface area and volume for them.

Directions: Answer the question below.

1. What are the units of measure for volume and surface area?

   A. Volume: units^3, Surface Area: units^3

   B. Volume: units^3, Surface Area: units^2

   C. Volume: units^2, Surface Area: units^3

   D. Volume: units^2, Surface Area: units^2

Directions: A box shaped like a rectangular prism has a width of 8.5 inches, a length of 9.5 inches, and a height of 14 inches. Use the information to answer questions 2 and 3.

2. What is the volume of the box?

   A. 252 in.^{3}

   B. 565.25 in.^{3}

   C. 1,130.5 in.^{3}

   D. 2,261 in.^{3}

3. What is the surface area of the box?

   A. 399.5 in.^{2}

   B. 427.5 in.^{2}

   C. 504 in.^{2}

   D. 665.5 in.^{2}

Directions: A concrete pad is being poured with dimensions of 50 feet long, 16 feet wide, and 1 foot tall. Use the information to answer questions 4 and 5.

4. What is the volume of the concrete that is being poured?

   A. 800 ft^3

   B. 1,600 ft^3

   C. 4,800 ft^3

   D. 9,600 ft^3

5. There is a glaze that needs to go on all visible surfaces of the concrete pad. What is the surface area of the glazed part?

A. 866 ft^2

B. 932 ft^2

C. 1,732 ft^2

D. 1,864 ft^2

6. The volume of a container shaped like a rectangular prism is 864 mm^3. It has a width of 4 mm and a length of 12 mm. The height of the container is _____.

Cylinders and Prisms

Prisms and cylinders both have heights, but the base of a prism is a polygon while the base of a cylinder is a circle.

Directions: A large can of food has a diameter of 8 centimeters and a height of 20 centimeters. Use the information to answer questions 7 and 8. Round each answer to the nearest tenth. Use 3.14 for π.

7. What is the volume of the can?

A. 507.4 cm^3

B. 1,004.8 cm^3

C. 2,009.6 cm^3

D. 4,019.2 cm^3

8. What is the surface area of the can?

A. 552.6 cm^2

B. 602.9 cm^2

C. 703.4 cm^2

D. 904.3 cm^2

Directions: A right triangular prism has sides that are 8 inches, 15 inches, and 17 inches long. It has a height of 20 inches. Use this information to answer questions 9 and 10.

9. _____ is the volume of the prism.

10. The surface area is _____.

Directions: A tent is shaped like a triangular prism with a length of 4 feet. The front and rear tent flaps are shaped like triangles, each with a base of 3 feet, a height of 2 feet, and two side lengths of 2.5 feet. Use this information to answer questions 11 and 12.

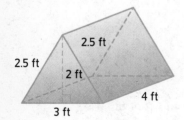

11. What is the volume of the tent?

A. 3 ft^3

B. 6 ft^3

C. 12 ft^3

D. 18 ft^3

12. What is the surface area of the tent?

A. 18 ft^2

B. 38 ft^2

C. 42 ft^2

D. 54 ft^2

Pyramids, Cones, and Spheres

Pyramids and cones are both solids with a height and a vertex, but the base of a pyramid is a polygon and the base of a cone is a circle.

Directions: A square pyramid has a base with side lengths of 12 centimeters and a height of 8 centimeters. Use this information to answer questions 13 and 14.

13. What is the volume of the pyramid?

    A. 96 cm³

    B. 144 cm³

    C. 288 cm³

    D. 384 cm³

14. The surface area of the square pyramid is _____.

    A. 336 cm²

    B. 384 cm²

    C. 528 cm²

    D. 624 cm²

Directions: A giant inflatable ice cream cone has a diameter of 10 inches and a height of 12 inches. Use this information to answer questions 15 and 16. Round your answers to the nearest tenth, when applicable. Use 3.14 for π.

15. The volume of the cone is _____.

    A. 314 in.³

    B. 942 in.³

    C. 1,256 in.³

    D. 1,884 in.³

16. What is the surface area of the cone?

    A. 219.8 in.²

    B. 266.9 in.²

    C. 282.6 in.²

    D. 690.8 in.²

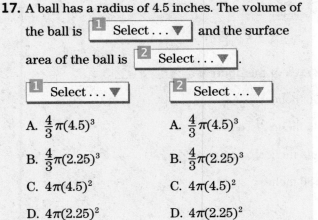

17. A ball has a radius of 4.5 inches. The volume of the ball is [1 Select . . . ▼] and the surface area of the ball is [2 Select . . . ▼].

[1 Select . . . ▼]

    A. $\frac{4}{3}\pi(4.5)^3$

    B. $\frac{4}{3}\pi(2.25)^3$

    C. $4\pi(4.5)^2$

    D. $4\pi(2.25)^2$

[2 Select . . . ▼]

    A. $\frac{4}{3}\pi(4.5)^3$

    B. $\frac{4}{3}\pi(2.25)^3$

    C. $4\pi(4.5)^2$

    D. $4\pi(2.25)^2$

18. A sand pile is conical in shape and has a volume of 9,800 cubic feet. The diameter of the pile is 40 feet. The height, to the nearest foot, is _____.

Test-Taking Tips

When answering questions related to surface area and volume of cylinders, cones, and spheres, you should double-check whether the radius or diameter is given in the problem. Recall that the radius is used in the formulas for all surface area and volume problems dealing with these three types of three-dimensional figures.

This lesson will help you analyze dimensions of composite figures by breaking them into their component shapes. Use it with core lesson 7.4 *Compute Perimeter, Area, Surface Area, and Volume of Composite Figures* to reinforce and apply your knowledge.

Key Concept	Core Skills & Practices
To find the area of a composite figure, add the area of each figure in the composite. To find the perimeter, add pieces of the perimeter of each figure. Similarly, to find the volume of a composite solid, add the volume of each solid. To find the surface area, add parts of each solid's surface area.	• Calculate Area • Calculate Surface Area of 3-Dimensional Solids

2-Dimensional Figures

In the real world, you may need to find the area and perimeter of two-dimensional composite shapes, such as the floor plan of an apartment or an odd-shaped backyard.

Directions: Use the diagram below to answer questions 1 and 2.

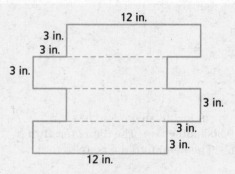

1. The perimeter of the figure is _____.

   A. 45 inches

   B. 48 inches

   C. 54 inches

   D. 66 inches

2. The area of the figure is _____.

   A. 36 inches²

   B. 45 inches²

   C. 144 inches²

   D. 180 inches²

 Test-Taking Tip

When calculating the area and perimeter of composite shapes, first recognize each shape that makes up the figure. Then, write down any formulas you need to use so you can keep your calculations organized.

Volume of 3-Dimensional Solids

Knowing how to calculate the volume of a composite figure is useful whenever you need to find the capacity of something made up of more than one shape.

Directions: Answer the questions below.

3. Find the volume of the figure below to the nearest mm³.

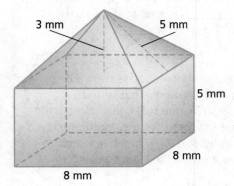

3 mm   5 mm

5 mm

8 mm

8 mm

A. 324 mm³

B. 341 mm³

C. 384 mm³

D. 405 mm³

4. A square pyramid with a height of 8 centimeters is stacked on a cube. The side of the cube is $2\frac{1}{2}$ times greater than the height of the pyramid. What is the volume of the composite figure to the nearest cubic centimeter?

A. 8,067 cm³

B. 9,067 cm³

C. 11,200 cm³

D. 16,000 cm³

5. A vitamin capsule has the shape of a cylinder with a hemisphere on each end. The radius of each hemisphere is 2mm and the height of the cylinder part of the capsule is 6mm. Find the volume of the capsule, and round to the nearest mm³.

A. 92 mm³

B. 109 mm³

C. 142 mm³

D. 159 mm³

6. A farmer owns a silo that is shaped like a cylinder with a hemisphere on top. The cylinder part of the silo is 45 feet tall and the height of the hemisphere is $\frac{1}{3}$ the height of the silo. The farmer can store _____ cubic feet of grain in the silo. Use 3.14 as pi and round your answer to the nearest whole number.

7. A perfume bottle is packaged in a box shaped like a cone sitting on top of a cylinder with a radius of 5 cm. The height of the cylinder is 24 cm and the height of the cone is $\frac{1}{2}$ the height of the cylinder. The volume of the box rounded to the nearest whole number is ☐ Select . . . ▼ .

☐ Select . . . ▼

A. 439 cm³

B. 753 cm³

C. 2,199 cm³

D. 3,770 cm³

Surface Area of 3-Dimensional Solids

You may need to calculate surface area when you do real-world tasks such as painting a house or wrapping a gift.

Directions: Answer the questions below.

8. Find the surface area of the figure below.

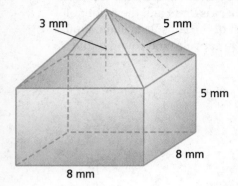

3 mm   5 mm

5 mm

8 mm

8 mm

A. 240 mm²

B. 304 mm²

C. 412 mm²

D. 444 mm²

9. Use the information from the silo problem in question 6. Now the farmer wants to paint the exterior of the silo. If a gallon of paint covers 400 square feet, the farmer needs _____ gallons of paint.

10. Use the information from the capsule problem in question 5. The vitamin manufacturer puts a coating around each capsule to make it easier to swallow. The surface area of the coating rounded to the nearest whole number is

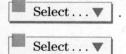 Select . . . ▼ .

■ Select . . . ▼

A. 126 mm²

B. 151 mm²

C. 176 mm²

D. 201 mm²

11. Use the information from question 7 on the previous page. The surface area of the perfume bottle box is _____. Round to the nearest whole number.

12. A sports company has decided to make basketballs for a local women's basketball league. According to league rules, the basketball must hold at least 121.5π in.³ of air. Assuming the ball is perfectly spherical, how much material must be used to adhere to the league rules?

A. 4.5π square inches

B. 9π square inches

C. 40.5π square inches

D. 81π square inches

13. Write the following shapes in order from least surface area to greatest surface area.

Sphere with radius 5	Cube with side length 5
Cylinder with radius 5 and height 6	Square pyramid with base length 5 and slant height 6

Least Surface Area

Greatest Surface Area

This lesson will help you describe and summarize sets of numbers, and determine common measures of tendency, including mean, median, mode, and range. Use it with core lesson 8.1 *Calculate Measures of Central Tendency* to reinforce and apply your knowledge.

Key Concept

A measure of central tendency is a number that can be used to summarize a group of numbers. Mean, median, and mode are measures of central tendency calculated in different ways.

Core Skills & Practices

• Interpret Data Displays

Measures of Central Tendency

The measure of central tendency that may be the most familiar is the mean, or average. But other measures can be more useful in certain situations.

Directions: Use the data set shown below to answer the questions 1–5.

32	31	33	44	35
44	37	44	40	41
38	38	40	32	44

1. What is the mean of the data set?

   A. 44

   B. 38.2

   C. 35.8

   D. 13

2. What is the mode of the data set?

   A. 44

   B. 38.2

   C. 38

   D. 12

3. What is the median of the data set?

   A. 44

   B. 38.2

   C. 38

   D. 13

4. The 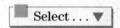 Select . . . ▼ of the data set is 13.

   Select . . . ▼

   A. mean

   B. median

   C. mode

   D. range

5. If another data entry, 60, is added to this set, the  Select . . . ▼ is now 39.6.

   Select . . . ▼

   A. mean

   B. median

   C. mode

   D. range

Finding a Missing Data Item

Sometimes an average is a goal you want to meet, and you'll need to find a missing piece of data to see if you can achieve your goal.

Directions: Answer the questions below.

6. In a science class, Amy wants to earn an 84% or better. The grade for the class is an average of the four tests taken. On the first three tests, she scored a 78%, 82%, and 86%. What is the lowest percentage she can make on the fourth test and still meet her goal?

A. 78%

B. 80%

C. 88%

D. 90%

7. Dontell sells software for a computer company. If he sells a weekly average of at least $1,000 in new software over an 8-week period, he will receive a bonus. His weekly sales so far are shown in this chart. How much does Dontell need to sell in Week 8 to receive the bonus?

Dontell's Weekly Sales	
Week	Sales
Week 1	$900
Week 2	$1,500
Week 3	$785
Week 4	$895
Week 5	$973
Week 6	$1,100
Week 7	$875

A. $715

B. $785

C. $972

D. $7,028

8. Allison wants to earn a 90% in her History class. The grade for the class is an average of the grades on four tests. She earns an 87%, 85%, and 92% on three tests. Which equation can she use to find out whether her goal is achievable?

A. $\dfrac{97 + 85 + 92 + 90}{3} = x$

B. $\dfrac{87 + 85 + 92 + x}{4} = 90$

C. $\dfrac{87 + 85 + 92 + x}{3} = 90$

D. $\dfrac{87 + 85 + 92 + 90}{4} = x$

9. Wes wants to earn a 95% or better in his History class. The grade for the class is an average of the grades on six tests. On the first five tests, Wes scored 68%, 89%, 92%, 75%, and 80%. Can Wes reach his goal if the next test doesn't have any extra credit on it?

_____

10. Karina's office is open for 10 hours a day, Monday through Friday. She worked 4 hours on Monday, 6 hours on Tuesday, 8 hours on Wednesday, and 7 hours on Thursday. Karina wants to work an average of 9 hours a day for the work week. She plans to work on Friday. Is it possible for Karina to meet her goal?

_____

 Test-Taking Tip

When finding a missing data item, make sure you set up your equation correctly, with the variable on the appropriate side. Then make sure you correctly use the order of operations to solve for the missing data item. Be sure to check that your answer makes sense in the context of the problem.

Weighted Averages

In some situations, a grade or other piece of data counts more than others, so it's important to know how to find an average (mean) when items are not all worth the same.

Directions: Shoppers at a mall were asked how many pairs of jeans they owned. The results are shown in the frequency chart below. Use the chart to answer questions 12–13.

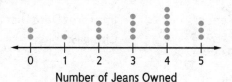

Number of Jeans Owned

11. The mean number of jeans owned by those surveyed is _____.

12. Suppose two more shoppers are surveyed. _____ is the number of jeans they would each need to own so that the average number of pairs of jeans owned is 2.75.

Directions: Answer the questions below.

13. Tanya is a photographer. She has 30-minute, 60-minute, and 90-minute sessions available. She charges $75 for a 30-minute session, and has 25 booked this month. She charges $150 for a 60-minute session, and has 30 booked this month. She charges $200 for a 90-minute session, and has 15 booked this month. Which is the average price Tanya will charge for each session this month?

   A. $52.08

   B. $133.93

   C. $162.50

   D. $3,125.00

14. A bakery sells 30 cupcakes at $3 each, 12 cakes at $20 each, and 40 loaves of bread at $4 each. What is the average price of an item sold?

   A. $5.98

   B. $9.00

   C. $18.15

   D. $490

15. The final grade in Miguel's chemistry class is based on 3 tests and a final exam. The final exam is worth 3 times as much as a test. Miguel earned the following test grades: 87, 82, 94. He earned a 91 on the final exam. Which is Miguel's final grade in the class?

   A. 87.0

   B. 88.5

   C. 89.3

   D. 94.0

This lesson will help you understand how to summarize information about different categories using bar graphs and circle graphs. Use it with core lesson 8.2 *Display Categorical Data* to reinforce and apply your knowledge.

Key Concept

Bar graphs and circle graphs are convenient ways of displaying data that fall into categories. Both types of graphs allow the viewer to see data at a glance. Bar graphs are appropriate to show the absolute size of various categories. Circle graphs show what percentage of the total is made up by the various categories.

Core Skills & Practices

- Interpret Data Displays
- Circle Graphs

Bar Graphs

In a bar graph, the relative length or height of the bars shows the relative size of the different categories.

Directions: Answer the questions below. Use the bar graph below for questions 1 and 2.

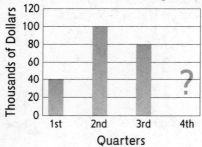

Chandra's Quarterly Sales

1. Chandra recorded her quarterly sales in a bar graph. Chandra will get a bonus of $5,000 if her average quarterly sales for the year reach $70,000. What must her sales be for the fourth quarter in order for Chandra to earn the bonus?

   A. $45,000

   B. $50,000

   C. $55,000

   D. $60,000

2. What was the percent of increase of the second quarter over the first quarter?

   A. 60%

   B. 100%

   C. 150%

   D. 200%

Directions: The bar graph shows the amount of interest paid on a $100,000 mortgage for two different pay-back periods and for three different yearly interest rates. Use the graph for questions 3 through 5.

Total Interest Paid on a $100,000 Mortgage

Legend:
- 6%
- 9%
- 12%

Y-axis: Thousands of Dollars (0 to 300, in intervals of 20)
X-axis: Mortgage Type (15-year, 30-year)

3. Suppose you want to borrow $100,000 to purchase a house, and the interest rate of your loan is 12%.

   Estimate the difference in total interest you would pay if you take out a 30-year loan instead of a 15-year loan.

   A. about $75,000

   B. about $150,000

   C. about $200,000

   D. about $250,000

4. What total amount of interest would you pay on a 30-year $100,000 mortgage at an interest rate of 9%?

   A. about $50,000

   B. about $80,000

   C. about $118,000

   D. about $190,000

5. Over the life of a 30-year mortgage at 6%, _____ is the average interest paid per year for a mortgage of $100,000. Round the average interest to the nearest $1,000.

 Test-Taking Tips

Look carefully at any notations beside the vertical scale when reading a bar graph. If there is a notation reading "Thousands of Dollars," for instance, a reading on the scale that says "$100" is to be interpreted as $100,000.

6. You want to create a bar graph displaying data in a range from $0 to $1000. What would be the best interval between ticks on the vertical scale?

   A. $5

   B. $100

   C. $500

   D. $1,000

Circle Graphs

Directions: Election results for a mayoral race are summarized in the circle graph. Use the graph for questions 7 through 10.

Mayoral Election Results: 50,200 votes cast

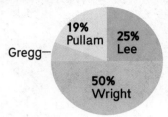

7. Gregg received _____ percent of the votes.

8. _____ received about $\frac{1}{5}$ of the votes cast.

9. What percent of registered voters voted in this election?

   A. between 30% and 60%

   B. between 60% and 80%

   C. more than 80%

   D. Not enough information is given.

10. According to information provided by the graph, what number of votes did Lee receive?

   A. 6,245

   B. 8,050

   C. 9,500

   D. 12,550

11. On the basis of the circle graph, write the name of each category next to the percent of the graph that it occupies.

   5% _____

   20% _____

   30% _____

   45% _____

   Preferred Mode of Transportation

   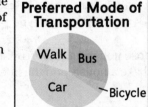

Directions: Use the graphs below for questions 12 and 13.

Mike's Bicycle Store
(Profit by Quarter)

Mike's Bicycle Store
Percent of Total Profit, by Quarter

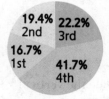

Total Profit = $18,000

12. The profit for Mike's Bicycle Store for the four quarters of the year is shown. What is the store's mean (average) profit per quarter?

   A. $4,500

   B. $5,600

   C. $6,250

   D. $6,800

13. Which expression tells the dollar amount of profit made during the first quarter?

   A. 1.67 × $15,000

   B. 16.7 × $18,000

   C. 0.167 × $15,000

   D. 0.167 × $18,000

This lesson will help you display data in different ways (dot plots, histograms, box plots) to highlight certain aspects of the data. Use it with core lesson 8.3 *Display One-Variable Data* to reinforce and apply your knowledge.

Key Concept

Dot plots, histograms, and box plots are different ways to display one-variable data, data in which only one quantity is measured. Each display highlights different characteristics of the data set.

Core Skills & Practices

- Interpreting Data Displays
- Model with Mathematics

Dot Plots

You can use a dot plot to find the mean, median, and mode of the data set.

Directions: Cassie surveyed students in her class and asked how many computers each person has at home. Her data are displayed below. Use the dot plot to answer questions 1–5.

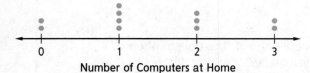

Number of Computers at Home

1. How many people did Cassie survey?

   A. 9

   B. 11

   C. 16

   D. 18

2. What is the median number of computers in a household?

   A. 0

   B. 1

   C. 2

   D. 3

3. What fraction of students surveyed own at least 2 computers at home?

   A. $\frac{1}{11}$

   B. $\frac{3}{11}$

   C. $\frac{5}{11}$

   D. $\frac{10}{11}$

4. The data set for this dot plot, from least to greatest, is _____.

5. Cassie surveyed an additional person and found that there were 3 computers at home. Order the new categories from least to greatest.

   0 computers | Least |

   1 computer | |

   3 computers | |

                  | Greatest |

Histograms

While histograms look like bar graphs, they are different because the data is numerical instead of categorical.

Directions: A local hospital recorded the ages of first-time mothers and displayed the data below. Use the histogram to answer questions 6–10.

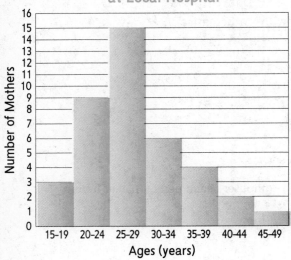

6. How many first-time mothers are in the data set?

A. 1

B. 7

C. 15

D. 40

7. What is the range of each interval?

A. 4

B. 5

C. 7

D. 15

8. How many first-time mothers are older than 34?

A. 1

B. 5

C. 7

D. 13

9. What percentage of first-time mothers are younger than 30?

A. 27%

B. 32.5%

C. 55%

D. 67.5%

10. A possible data set for this histogram, in order from least to greatest, is _____

_____

_____

_____

_____ .

Test-Taking Tips

When interpreting a data display, remember that not all displays show the same information. Be sure to read labels carefully so that you can confirm what is actually conveyed in the data display. If the question requires listing a set of data, you can check your work by sketching a data display to see if it matches the one given in the problem.

Box Plots

A box plot can help you understand how spread out a set of data is.

Directions: Henry took a survey of his friends to find out how much money they make per hour. He displayed the results in the box plot below. Use the box plot to answer questions 11–17.

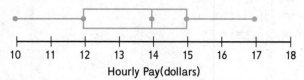

Hourly Pay(dollars)

11. For the box plot above, the median is _____, 1st quartile is _____, and the 3rd quartile is _____.

12. A possible data set for this box plot is

_____

_____ .

13. What is the median hourly pay among Henry's friends?

   A. $10

   B. $12

   C. $14

   D. $17

14. What is the range of the data?

   A. 7

   B. 10

   C. 14

   D. 17

15. Based on the box plot, which of the following statements is true?

   A. One quarter of Henry's friends make between $10 and $14 an hour.

   B. Half of Henry's friends make between $15 and $17 an hour.

   C. One quarter of Henry's friends make between $14 and $17 an hour.

   D. Half of Henry's friends make between $12 and $15 an hour.

16. A data set contains the following items:

25, 30, 40, 42, 9, 9, 12, 10, 15, 25, 35, 41, 45, 10, 12, 25, 26.

The data can be displayed in a box plot. The least value is [1] Select . . . ▼ and the greatest value is [2] Select . . . ▼ . The left side of the box is the first quartile, 11, and the third quartile is [3] Select . . . ▼ . The line through the box is drawn at the [4] Select . . . ▼ .

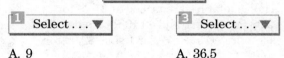

[1] Select . . . ▼

   A. 9

   B. 10

   C. 11

   D. 25

[3] Select . . . ▼

   A. 36.5

   B. 37

   C. 37.5

   D. 38

[2] Select . . . ▼

   A. 26

   B. 40

   C. 45

   D. 46

[4] Select . . . ▼

   A. mean, 25

   B. mean, 26

   C. median, 25

   D. median, 26

17. The following data set is missing a value. To make a box-and-whisker plot, Ed draws whiskers from 4 to 8 and from 16 to 20. Which could be the missing data?

8, 13, 20, 14, 6, 12, 18, 4, 11, 16

   A. 4

   B. 10

   C. 17

   D. 21

This lesson will help you analyze two-variable data using tables, scatter plots, and line graphs. Use it with core lesson 8.4 *Display Two-Variable Data* to reinforce and apply your knowledge.

Key Concept

Tables, scatter plots, and line graphs are all ways to show information that relates one thing to another, like temperature to time of day or height to weight. We call these displays of two-variable data, because there are two items that have a relationship.

Core Skills & Practices

• Build Lines of Reasoning
• Interpret Graphs

Tables

Tables can organize and display a wide array of data including prices of items at a restaurant, sports statistics, and populations of cities.

Directions: Answer the questions below.

1. Create the table for the following information: At a restaurant, there are various prices for side dishes. The vegetables are priced at $1.29 for 1 side, $2.39 for 2 sides, and $3.19 for 3 sides. The potatoes are priced at $1.79 for 1 side, $2.99 for 2 sides, and $3.89 for 3 sides. The fruits are priced at $1.49 for 1 side, $2.89 for 2 sides, and $4.19 for 3 sides. Fill in the missing information.

	1 side	2 sides	3 sides
Vegetables			
Potatoes			
Fruits			

2. A company makes tank tops, t-shirts, and long-sleeve shirts by 4 different designers. How large a data table does the company need to organize the prices of all the different shirts?

_____columns _____ rows

Directions: The table below represents the number of ants, bacteria, and birds in an environmental setting. Use the table to answer questions 3 and 4.

	Week 1	Week 2	Week 3	Week 4
Ants	50	147	268	319
Bacteria	8	201	472	981
Birds	125	119	108	102

3. During week 3, there were _____ more bacteria than birds.

A. 82

B. 117

C. 209

D. 364

4. The week(s) that had more bacteria than ants and birds combined were _____.

A. Weeks 1 and 2

B. Week 2

C. Weeks 3 and 4

D. Week 4

Scatter Plots

Scatter plots are used to plot distinct points relating two variables, whether or not a relationship may actually exist between the variables.

Directions: Create a scatter plot using the data table below.

5. The table shows height and weight statistics for basketball players on a team. Use the blank graph to create a scatter plot from the data in the table, by plotting each data point.

Height (inches)	83	80	77	72	74	76	84	75	82	78
Weight (pounds)	250	230	210	190	170	170	200	220	240	180

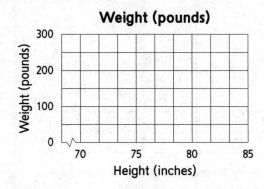

Directions: Use the following table to answer question 6. The table represents data from a shopping trip by 12 different people.

Hours Shopping	5	6	6	7	8	9	9	10	11	11	12	12
Dollars Spent	300	417	427	518	529	681	724	851	902	988	1016	1103

6. Create a scatter plot from the data in the table, by plotting each data point.

✔ Test-Taking Tip

When plotting points on a scatter plot, be sure to look at the scale of the axes so you are plotting points in the correct locations.

Line Graphs

Line graphs are useful ways to display data that can increase or decrease at any time, such as company profits or the temperature outside.

Directions: Answer the question below.

7. Which of the following would be most appropriate to display on a line graph?

   A. List of prices of two brands of cars and trucks

   B. Hours spent studying and test scores earned by students in a class

   C. Number of products sold each day during the course of a month

   D. Inches of rainfall each month over the course of a year

Directions: The table below shows a city's average temperature in Celsius for certain years. Use the table to answer questions 8–11.

Year	1998	2000	2002	2004	2006	2008	2010	2012
Temperature	8	6	4	12	9	5	1	7

8. Create a line graph for the data by plotting each point.

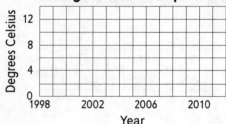

9. What is the trend that can be concluded from the graph?

   A. The average temperature increases every year.

   B. The average temperature decreases every year.

   C. There is an increase in temperature one year, and then a decrease the next year.

   D. There is no trend between year and temperature.

10. Which two years' average temperatures add up to the average temperature of 2004?

    A. 2006 and 2012

    B. 2000 and 2002

    C. 2008 and 2012

    D. 2000 and 2008

11. The year 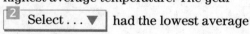 had the highest average temperature. The year 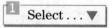 had the lowest average temperature.

1 Select . . . ▼	2 Select . . . ▼
A. 1998	A. 2000
B. 2004	B. 2006
C. 2006	C. 2008
D. 2012	D. 2010

Lesson 1.1

Rational Numbers, p. 1

1. **B**   Looking at the number line, the numbers that are represented are −3, 0.5, and 2.

2. **C**   The whole numbers are the positive counting numbers as well as zero.

3. **C**   Rational numbers can be written as a ratio of two integers. Therefore, irrational numbers are numbers that cannot be written as the ratio of two integers.

4. **B**   The square root of 1 equals 1, which is a rational number.

Fractions and Decimals, p. 2

5. **D**   The number 5 goes into 28 5 times with 3 as a remainder.

6.

Shortest $2\frac{3}{12}$	$2\frac{3}{8}$	$2\frac{6}{10}$	Longest $2\frac{3}{4}$

7. $0.125 \underline{\quad < \quad} \frac{1}{6}$

8. **C**   The fractional part is $\frac{3}{4}$, which when divided out, equals 0.75. The fraction equals 5.75.

9. **C**   All 4 numbers agree on the first two digits on the left. The first digit to the right of the decimal point has differences, with 5 being the largest.

Absolute Value, p. 3

10. **A**   Moving 3 units to the left of 2 results in the number −1. Moving 3 units to the right of 2 results in the number 5.

11. **C**   The distance between −6 and 3 can be found by finding |−6 − 3| = |−9| = 9.

12. The absolute value of the sum of −17 + 8 is __9__ .

13.

Least $|-1|$	6	$|-12|$	19	Greatest $|-20|$

14. **D**   The distance between −4 and −6 can be found by taking the absolute value of their difference.

15. **D**   The number of bags on back order is 538 − 217 = 321.

Lesson 1.2

Factors and Multiples, p. 4

1.

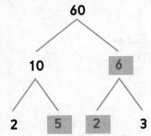

2. $60 = 2^{\boxed{2}} \times \boxed{3} \times \boxed{5}$

3. **C**   Since each group must have the same number of distance runners and sprinters, the number of groups must be a factor of both the total number of distance runners and sprinters. The largest number that is a factor of both 20 and 25 is the GCF of the two numbers, 5.

Answer Key

4. **B** Each collar requires 1 buckle and 1 strap. The total number of straps and buckles needed must be a multiple of both 12 and 16. 48 is the LCM of 12 and 16, and so dividing 48 by 12 and 16 gives 4 packs of buckles and 3 packs of straps.

5. **D** The side length of the grid squares must divide evenly into both side lengths. Therefore, the length of one grid square must be a factor of both 12 and 30, with 6 being the GCF.

Properties of Numbers, p. 5

6. **A** The expression is an example of the Distributive Property. Distributing the 2 outside the parenthesis inside will result in $2(6 + 3) = 2 \times 6 + 2 \times 3$.

7. **D** The Commutative Property of Addition states that two numbers can be added in any order and the result is the same.

8.

Example	Property
$(2 \times 3) \times 4 = 2 \times (3 \times 4)$	Associative Property of Multiplication
$7 + (10 + 3) = 7 + (3 + 10)$	Commutative Property of Addition
$4(a + 6) = 4a + 24$	Distributive Property
$6 \times 7 = 7 \times 6$	Commutative Property of Multiplication

9.

| Select . . . ▼ |

C. equal to

Order or Operations, p. 6

10. The numbers __7__ and __−7__ would make the following expression undefined.

$20 \div (49 - m^2)$

11. If 100 is the best possible score on a test and any score over 70 is passing, is a score of $100 - \frac{10}{2} \times 20$ a passing score? __No__

12. **D** Kevin divided 16 by 8 as well as took the cube of 6, ignoring the parentheses.

13. **A** Marcie spends $15.00 on 3 shirts, or 3×15 dollars. She adds a $30.00 jacket, which is $5.00 off, or $(30 - 5)$ dollars. The $10.00 gift card subtracts $10.00 from her total, which is $(3 \times 15 + (30 - 5))$.

14. **C** Using the Order of Operations, you can evaluate

$5 + \frac{(6^2 - 10)}{2} + 3 = 5 + \frac{(36 - 10)}{2} + 3 = 5 + \frac{26}{2} + 3 = 5 + 13 + 3 = 21.$

Lesson 1.3

Exponential Notation, p. 7

1. **D** The total area of the room is $14 \times 14 = 14^2$ square feet. At $9.50 per square foot, that cost is 9.50×14^2 dollars. Adding the installation fee gives the expression $9.50 \times 14^2 + 75$ dollars.

2. What is the value of the expression $x^2 + y^3$ for $x = 4$ and $y = 2$? __24__

3.

Least Money		Most Money
Isaiah	Matt	Sam

4. **A** The volume of the large cubic container is 4^3 cubic inches and the volume of the small cubic container is 3^3 cubic inches. With 6 large cubic containers, the total volume is $6 \times 4^3 + 3^3$.

5.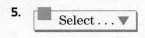

     D. 14

Rules of Exponents, p. 8

6. C   The expression $(4a^4)^2$ has a power raised to a power during simplification.

7. $(t^{\underline{-3}})^4 = t^{-12}$

8.

Least		Most
$12^5 \div 12^2$	$12^2 \times 12^5$	$(12^2)^5$

9. A   Simplifying the numerator using the Quotient of Powers Property reduces the expression to $\dfrac{4^{-2+4}}{4^2} = \dfrac{4^2}{4^2} = 1$.

10. D   Using the Power of a Power Property simplifies the expression to $2^{2 \times 4} \cdot 3^{3 \times 4}$. This simplifies to $2^8 \cdot 3^{12}$.

Scientific Notation, p. 9

11. D   A number written in scientific notation must be a number greater than or equal to 1 and less than 10 times a power of 10. To fix the mistake, 65.2 must be written as 6.52 with another power of ten added to the current power.

12. B   The number being multiplied to the power of 10 will be 5.04 since 504 is all of the non-zero portion of the number. Since the decimal place is being moved 8 places to the right, the power of ten will be -8.

13. She can grow $\underline{6.453 \times 10^9}$ bacteria in 10 Petri dishes, expressed in scientific notation.

14. C   To find the sum of numbers in scientific notation, write both numbers in the same power of 10. Rewriting $4.2 \times 10^5 = 420 \times 10^3$, you can then add 6.7×10^3 to get $426.7 \times 10^3 = 4.26 \times 10^5$.

15. B   After 5 years, the population will be $2(3.2 \times 10^7)$, or 64,000,000. Doubling that number gives the population after 10 years, 128,000,000. This number falls in the range 120 million to 140 million people.

Lesson 1.4

Square Roots and Cube Roots, p. 10

1. The __cube__ root of 27 is 3.

2. A   The area of the old plates is $10 \times 10 = 100$ square inches. The new plates should hold 1.5 times that or 150 square inches. The side length of the new square plate will satisfy $s^2 = 150$, so $s \approx 12.2$ inches.

3. C   The area of the square floor is 350 square feet. To find the side length, solve the equation $s^2 = 350$, which gives $s \approx 18.7$ feet.

4. C   The volume of the pool that holds 10,000 gallons, in cubic feet, is $\dfrac{10,000}{7.48} \approx 1337$ cubic feet. To find the side length of the cubical pool, solve the equation $s^3 = 1337$, which gives $s \approx 11.02$. Therefore, a cubical pool with side length 12 feet will hold 10,000 gallons.

5. The square root of an integer is either an integer or a(n) __irrational number__.

6. B   Taking an even root of a negative number results in a non-real answer. Only taking an odd root of a negative number results in a negative number.

7.

Least				Most
$\sqrt{140}$	12	$\sqrt{156}$	13	$\sqrt{170}$

8.

Perfect Square	Perfect Cube
1	1
64	8
100	27
144	64

9. 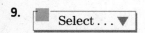 Select . . . ▼

   A. cube root

10. C To find the side length of the square, solve the equation $s^2 = 56.25$, which gives $s = 7.5$. The amount of fence required to enclose the area is $P = 4s = 30$ feet.

Radicals and Rational Exponents, p. 11

11. C The distance the fly will travel is 4 meters. At a rate of $\sqrt{2}$ meters per minute, the fly will walk 4 meters in $\frac{4}{\sqrt{2}} \approx 2.8$ minutes.

12. D Evaluating all 4 choices finds that $9^{\frac{3}{2}} = 81^{\frac{3}{4}} = \sqrt{9^3} = 27$, while $\sqrt[3]{9^2} \approx 4.3$.

13. The simplified answer to the expression $\frac{2}{\sqrt{2}}$ is $\underline{\sqrt{2}}$.

14. A To find the side length of a cube given its volume, take the cube root of the volume. The area of one face of the cube will be the square of that number.

15. D To simplify the expression, first combine the expression using the properties of exponents to $\frac{\sqrt{96}}{\sqrt{12}} = \sqrt{\frac{96}{12}} = \sqrt{8}$. Then, factor out perfect squares to get the expression to simplify to $\sqrt{8} = 2\sqrt{2}$.

16. B To find the side length of a cube given the area of one of its side, take the square root of the area. Then cube the side length to find the volume.

17. The simplified answer to the expression $\frac{\sqrt[3]{108}\ \sqrt[3]{16}}{\sqrt{9}}$ is $\underline{4}$.

18. A Evaluating all 4 choices finds that $\frac{2\sqrt{10}\ \times \sqrt[3]{8}}{\sqrt{2}} = 5^{\frac{1}{2}} \times \frac{2}{2^{\frac{1}{2}}} = \sqrt{5} \times \frac{\sqrt{2}}{2} = \frac{\sqrt{40}\ \times \sqrt[3]{8}}{\sqrt{2}}$.

19. A Rewriting the expression using exponents helps simplify multiple roots. $\sqrt{\sqrt[6]{2}} = \left(2^{\frac{1}{6}}\right)^{\frac{1}{2}} = 2^{\frac{1}{6} \times \frac{1}{2}} = 2^{\frac{1}{12}} = \sqrt[12]{2}$.

20. The simplified answer to the expression $\frac{\sqrt{25}}{\sqrt[3]{125}}$ is $\underline{1}$.

Lesson 2.1

Ratios, p. 13

1. C The ratio of boys to girls is $\frac{500}{120}$, which can be simplified to $\frac{25}{6}$, by dividing both the numerator and denominator by 20.

2. D The total amount spent on screws is $0.48(7 + 5) = 0.48 \times 12 = 5.76$.

3. A The largest set of animals is the cows, with 24, the smallest being the pigs with 15. However, the next smallest group is chickens with 16. Therefore, the largest ratio is cows to chickens.

4. B The first car travels at 60 miles per hour, while the second car travels at 55 miles per hour.

5.

Least				Most
bananas	pears	apples	peaches	grapes

Proportions, p. 14

6. A   Megan set up the wrong ratio. Working 5 hours on Friday, she made $48. Working 3 hours on Saturday, she made x dollars. This proportion would be $\frac{48}{5} = \frac{x}{3}$.

7. C   Set up the proportion $\frac{500}{4} = \frac{s}{60}$, where 60 is the seconds per minute. Solving for s gives $30,000 = 4s$, or $s = 7,500$.

8. C   Set up the proportion $\frac{68.40}{20} = \frac{d}{5}$ to find out how much 5 boxes cost. Solving for d gives $342 = 20d$, or $d = 17.10$.

9. The exchange rate for U.S. to Chinese currency is 2 dollars to 12 yuan. If you had 883.14 Chinese yuan, you would have __147.19__ U.S. dollars.

10. $\dfrac{\boxed{800}}{\boxed{2}} = \dfrac{\boxed{1800}}{\boxed{g}}$ __4.5__ gallons

11. D   Set up the proportion $\frac{4}{10} = \frac{m}{2,500}$. Solving for m gives $10,000 = 10m$, or $m = 1,000$.

12. [■ Select . . . ▼]

    D. 767

Scale, p. 15

13. $\dfrac{15}{\boxed{1}} = \dfrac{d}{\boxed{2\frac{1}{2}}}$

    $d = \boxed{37.5}$ miles

14. B   Set up the proportion $\frac{5.5}{8} = \frac{s}{14}$. Solving for s gives $77 = 8s$, or $s = 9.625$.

15. C   The perimeter of Mike's garden is $8 \times 5 = 40$ feet. Set up the proportion $\frac{n}{40} = \frac{3}{2}$. Solving for n gives $120 = 2n$, or $n = 60$ feet.

16. A wall in a new house has a height of 17 feet and a width of 19 feet. In a photo of the house, the width of the wall is 0.5 feet. __1:38__ is the scale factor of the photo compared to the actual wall.

17. B   The base of the large triangle is 10 inches and the base of the small triangle is 6 inches. The scale factor is $\frac{10}{6} = \frac{5}{3}$.

18. A   A leg of the smaller triangle is 18 inches. Multiply 18 to the scale factor $\frac{5}{3}$ to get $18 \times \frac{5}{3} = 30$ inches.

Lesson 2.2

Percent of a Number, p. 16

1. [Sample Answer]

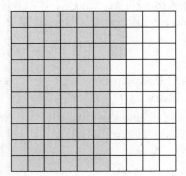

Answer Key

2. **C** Because 24% of people surveyed prefer running barefoot, 76% of people prefer running wearing a shoe. Multiplying $0.76 \times 300 = 228$ gives the number of runners that would rather run wearing a shoe.

3. **B** To find the percentage, divide 23 by 25 to get $\frac{23}{25} = 0.92$. Multiply this by 100 to get 92%.

4. **A** Multiplying $0.79 \times 3{,}000$ gives the number of people who would go into the water above their waist.

Percent Change, p. 17

5. **D** The percent change can be found by simplifying $\frac{126 - 180}{180} = -0.3$, which is a 30% discount.

6. year 1 and year 2 ___−14%___

   year 2 and year 3 ___4%___

   year 3 and year 4 ___20%___

   year 4 and year 5 ___−7%___

   year 5 and year 6 ___4%___

7. **B** After 20% was discounted, the new price was 80% of the old price. Subtracting the $20 it cost to make, multiply the whole expression by 200 to find the profit, or $200 \times ((0.8 \times 54.99) - 20)) = 200 \times (43.992 - 20) = 200 \times 23.992 = 4{,}798.4$, or approximately $4,800.

8. **C** Multiply the discount percentage as a decimal to the original price to find how much money was saved. The largest amount saved is $0.15 \times 36 = 5.40$ dollars.

9.

	Percent Change
Day 1 to Day 2	12%
Day 2 to Day 3	−4%
Day 3 to Day 4	6%
Day 4 to Day 5	−3%
Day 5 to Day 6	−4%
Day 6 to Day 7	6%
Day 1 to Day 7	13%

10. If the discounted price is $159.00, then the percent discount is ___25%___.

    If the discounted price is $125.08, then the percent discount is ___41%___.

    If the discounted price is $106.00, then the percent discount is ___50%___.

    If the discounted price is $99.64, then the percent discount is ___53%___.

Simple Interest, p. 18

11. **C** Using the simple interest formula, $I = Prt$, multiply the principal, rate, and time together to find the interest. Bank C will pay $I = 2{,}000 \times 0.0175 \times 8 = 280$ dollars.

12. The student will pay ___$840___ in simple interest on the loan in one year.

    The student will pay ___$12,600___ in simple interest on the loan over the life of the loan.

13. **C** Using the simple interest formula, $I = Prt$, multiply the principal, rate, and time together to find the interest. $I = 10{,}000 \times 0.0299 \times 4.5 = 1{,}345.50$. To find the total amount, add the interest to the principal.

14. The better deal for Sean is the ___5-year loan___.

15. **D** Use the simple interest formula to find the time it will take for Kayla to make $1,000 in interest. $1{,}000 = 1{,}000 \times 0.05 \times t$. Simplifying the equation gives $1{,}000 = 50t$, or $t = 20$ years.

16. **B** Divide each car loan amount by 36 (3 years) and use the simple interest formula to find the lowest interest per month. Car B has the lowest amount of interest per month since $7{,}500 \div 36 = 208.33$, and $208.33 \times 0.018 \times 1 = 3.75$ dollars.

Lesson 2.3

Factorials, p. 19

1. The number 5! equals ____120____.

2. The correct number of possible orders is ____12____.

3. **D** The number of ways to order 6 people in a line is 6!, or 720.

4. **C** There are 2 types of watches (men's and women's) with 3 sizes of wristbands. Each wristband can be made of 3 types of material in 3 different colors. The total number of watches is $2 \times 3 \times 3 \times 3 = 54$.

5. **D** There are 52 cards in a standard deck. With 2 choices for the coin and 4 choices for the spinner, there are $2 \times 52 \times 4 = 416$ choices.

6. The total number of possible orders from the Coffee Crew is ____12____.

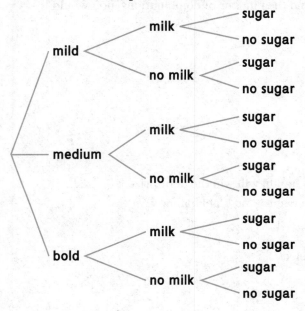

Permutations, p. 20

7. **A** The number of permutations of k objects taken from n objects is $P(n, k) = \dfrac{n!}{(n-k)!}$.

8. **B** When lining up 5 floats, the order matters and all floats are being used. Therefore, $P(5, 5) = \dfrac{5!}{(5-5)!} = \dfrac{5!}{0!} = \dfrac{120}{1} = 120$.

9. **B** Since numbers cannot be repeated and order matters, the number of possible combinations are $P(9, 3) = \dfrac{9!}{(9-3)!} = 9 \times 8 \times 7 = 504$ combinations. Trying one every 5 seconds, the longest it will take is $504 \times 5 = 2,520$ seconds or 42 minutes.

10.
    1 Select . . . ▼

    B. permutations

11. **C** The number of ways to cast the roles will be $P(8, 3) = \dfrac{8!}{(8-3)!} = 8 \times 7 \times 6 = 336$.

12. ____120 ways____

Combinations, p. 21

13. **B** The order of the speakers does not matter and so the total number of possibilities is $C(6, 3) = \dfrac{6!}{3!3!} = 20$.

14.

Permutations	Combinations
I, IV	II, III

15. A   There are 3 ways to order a pizza with 1 topping (mushroom only, onion only, and anchovy only), 3 ways to order a pizza with 2 toppings (mushroom and onion, mushroom and anchovy, and onion and anchovy), and 1 way to order a pizza with all 3 toppings. The total cost would be $3 \times 6.95 + 3 \times 7.95 + 1 \times 8.95 = 53.65$.

16. The difference between finding permutations and finding combinations is that when you are finding combinations, the _____order_____ of the items does not matter.

17. The correct number of combinations is ___10___.

18. C   The number of permutations is $P(27, 3) = \dfrac{27!}{(27-3)!}$. To find the number of combinations, you would need to divide by 3!, or 6.

Lesson 2.4

Probability of Simple Events, p. 22

1.

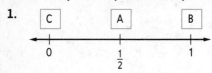

2. If 390 people attend, the probability that someone in Janine's family will win the door prize is ___$\frac{1}{130}$___.

3. C   There are 6 possible uniforms, with only 1 that has a green shirt and black pants.

4. Sal's guess is more like to be too ___high___.

5.

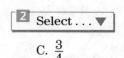

    B. complement   C. $\frac{3}{4}$

6. D   The probability of rolling a specific number is $\frac{1}{6}$. Therefore, you can expect to roll a specific number $30 \times \frac{1}{6} = 5$ times.

7.

Color	Number
Red	10
Yellow	7
Blue	8

Probability of Compound Events, p. 23

8. A   The probability of spinning a red section once is $\frac{2}{6} = \frac{1}{3}$. Spinning a red section twice is $\frac{1}{3} \times \frac{1}{3} = \frac{1}{9}$.

9. A   The probability of drawing one white marble on the first draw is $\frac{4}{9}$. Not replacing the marble leaves 8 marbles in the bag. Drawing a second white marble has probability of $\frac{3}{8}$. The total probability would be $\frac{4}{9} \times \frac{3}{8} = \frac{1}{6}$.

10. A   The probability of drawing a king is $\frac{1}{4}$ and the probability of flipping a head is $\frac{1}{2}$. The probability of doing both is $\frac{1}{4} \times \frac{1}{2} = \frac{1}{8}$.

11. C   The probability of drawing a queen on the first draw is $\frac{2}{4} = \frac{1}{2}$ and the probability of drawing a queen on the second draw is $\frac{1}{3}$ since the first queen was removed. Therefore, the probability of doing both is $\frac{1}{2} \times \frac{1}{3} = \frac{1}{6}$.

12. Drawing the second black is a(n) ___dependent___ event.

13. A   The probability of drawing a vowel twice is $\frac{6}{26} \times \frac{5}{25} = \frac{30}{650} = \frac{3}{65}$.

Lesson 3.1

Algebraic Expressions, p. 25

1. C   The expression $10p$ represents the total cost of food and $4p$ represents the total cost of beverages for p people. Adding in the \$350 rental fee, the expression that represents the total cost is $350 + 10p + 4p$.

2. A   The expression $-10t$ represents the number of feet the plane has descended after t seconds. Adding this to the original height of 50,000 feet, the expression that represents the descent is $50,000 - 10t$.

3. B   Using w as the width, the length is represented as $4w - 5$. The perimeter of the garden will then be $2(4w - 5) + 2w$.

4. D   When Robert works 40 hours in a week, he makes $12(40)$ dollars. For every hour of overtime (hours past 40), he makes 1.5 times he normally makes, which is $12(1.5)$ dollars per hour. Working h hours of overtime, he makes $12(1.5)h$ dollars for overtime.

5. Using t = time in months, an expression for the value of the truck over time is ___$40,000 - 300t$___

6. C   Cole takes n slices from 12 slices, leaving $12 - n$ slices for his 5 friends to share equally. Therefore, each friend receives $\frac{12 - n}{5}$ slices.

Linear Expressions, p. 26

7.

First Step
$(7x - 8) + 2(x - 5) = (7x - 8) + 2x - 10$

$(7x - 8) + 2x - 10 = 7x + 2x - 10 - 8$

Last Step
$7x + 2x - 10 - 8 = 9x - 18$

8. C   Distributing the -4 through the parenthesis would occur first.

9. $-2x -$ __27__ .

10. A   Distribute the -5 through the parenthesis to simplify the expression.

11. __10__ $x + 25$

12. D   Cassandra forgot to distribute the -1 to everything in the parenthesis.

13. C   The sum of the first two terms simplifies to $-4x - 8$. Changing the "-2" to "2" in the last term will add $2x - 28$ to the sum of the other two terms, which will result in $-2x - 36$ for the expression.

Evaluating Linear Expressions, p. 27

14. Evaluate the expression $3x + \frac{3}{2}y$ when $x = -4$ and $y = 2$. __-9__

15.

Less than 8	Equal to 8	Greater than 8
$2x + y$	$x - y$	$2x - 3y,\ 3x + y$

16. **1** Select . . . ▼   **2** Select . . . ▼

   B. \$31   A. \$29

Lesson 3.2

One-Step Equations, p. 28

1.

> **1** Select . . . ▼

> B. $n = 24$

2. $\underline{x - 8 = 31}$ is the equation that represents "8 less than x is 31".

3. D   There are 150 people being divided equally into s teams, so $150 \div s$ shows how many people per team.

4. B   In this scenario, the 6 lawns have no bearing on what is being asked. Paying $18 dollars for gas means that he had $n - 18$ dollars left, which is $90.

5. A   Add 10 to both sides of the equation to solve for w.

6. A   Since -56 needs to be moved to the other side of the equals sign, 56 should be added to both sides.

7. B   Divide both sides of the equation by -12 to find the solution to the equation.

Multi-Step Equations, p. 29

8. C   The phrase "5 plus the product of 4 and z" represents the expression $5 + 4z$, which is equal to 49. The equation $5 + 4z = 49$ can be solved by first subtracting 5 from both sides and then dividing by 4 on both sides.

9. C   First, subtract 27 from both sides of the equals sign, which gives $-6x = -60$. Then divide both sides by -6 to get $x = 10$.

10. A   Substitute 101 for j. This gives the equation $2b + 3 = 101$. Subtracting 3 from both sides gives $2b = 98$, which simplifies to $b = 49$ when you divide both sides by 2.

11. C   Dividing both sides of the equation by -7 gives the equation $m + 4 = -2$, which simplifies to $m = -6$. Therefore, $9m = 9 \times (-6) = -54$.

12. A   The phrase "the difference of a number and 5" represents the expression $n - 5$. This is then multiplied by -9 and set equal to 81, or $-9(n - 5) = 81$. Divide both sides of the equation by -9 results in $n - 5 = -9$, which simplifies to $n = -4$.

13. The cost of each breakfast before tipping is $\underline{\$11}$.

14. D   Simplifying the right side of the equation results in $-18 = -12n + 30$. Subtracting 30 from both sides of the equation gives $-48 = -12n$, or $n = 4$.

15. The value of n that makes the equation $12 + n = 7n + 2(n - 3)$ true is $\underline{2.25}$.

16. A   Adding 7 to both sides of the equation and then dividing by 5 gives the solution $y = 11$. Substituting that into the expression $6y + 4$ gives $6(11) + 4 = 66 + 4 = 70$.

17.

$x = 13$	$x = -7$
$-5x + 9 = -2x - 30$	$20x + 35 = 15x$
$x - 12 = 3(x - 10) - 8$	$-x + 13 = 4(x + 12)$
$-7(8 - x) = 3x - 4$	$5x - 1 = -9(x + 11)$

18. B   Distributing through the parenthesis and simplifying gives the equation $12s - 48 = 10s - 10$. Subtracting $10s$ from both sides and adding 48 to both sides gives $2s = 38$, or $s = 19$.

19. C   The equation that represents the scenario is $8 + 4m = 52$, where m is the number of model cars Johann will need to sell. Subtracting 8 from both sides and then dividing by 4 gives $m = 11$ cars.

Lesson 3.3

Inequalities, p. 31

1. B   The solution to the inequality is all numbers less than or equal to -4.

2. 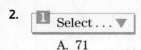   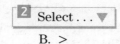   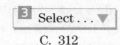

      A. 71   B. >   C. 312

3. C   The phrase "three times the sum of a number and seven" is represented by the expression $3(x + 7)$. This expression should be "greater than or equal to half of the number", which is why $\geq$ is the symbol that is used.

4. D   The inequality that is represented by the scenario is $2{,}400 < 8b$, where b is the number of bake sale items sold. Solving this results in $b > 300$.

One-Step Inequalities, p. 32

5. D   The numbers $-\frac{8}{3}$, $-2\frac{2}{3}$, and $-2.666\ldots$ all represent the same number.

6. A   Subtracting 12 from both sides of the inequality results in the solution $x < -3$.

7. B, C, D   Substitute -8 into each inequality; the inequalities B, C, and D are all true.

8. B   Multiplying both sides of the inequality by -7 gives the inequality $x < -7$.

9. D   The inequality represented by the scenario is $d - 400 \geq 1{,}000$, or $d \geq 1{,}400$.

Multi-Step Inequalities, p. 33

10. $p \leq \underline{\ 15\ }$

11. D   Simplify the inequality by using the Distributive Property on both sides of the inequality first. $28 - 7a + 1 < 1 - 4a - 20$. After simplifying, the inequality gives $29 - 7a < -19 - 4a$, or $48 < 3a$ after moving terms to the opposite side.

12. B   Subtract $\frac{2}{3}x$ from both sides of the inequality and add 8 to both sides of the inequality. After simplification, this results in $\frac{1}{12}x \geq 2$, or $x \geq 24$.

13. $x \geq \underline{\ 91\ }$

14. 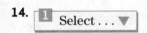   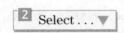

      D. distribute the 5 through the parenthesis   A. $8y \geq 2$

15.

$y > \dfrac{4}{3}$	$y < -\dfrac{4}{3}$
$-3y + 5 > -6y + 9$	$7y < y - 8$
$-y + 12 < 2(y + 4)$	$2y > 11(y + 1) + 1$
$-1 < 3y - 5$	$4 + 3(y - 1) < -3$

Lesson 3.4

Expressions and Equations, p. 34

1. D   For a ride of m miles, Nick charges $0.25m$ dollars. Adding in the \$4 flat fee creates the expression $4 + 0.25m$.

2. A   The expression $25n$ models how much money Freddy will be paid for mowing n lawns. Since he has saved \$35, he only needs to save \$$(110 - 35)$ more. Setting both expressions equal is the equation needed.

3. An equation to determine the average speed (s, in miles per hour) that she drove during her trip is

$$\frac{165}{3.25 - .25} = s.$$

4. B The equation $150 - (8h) = 54$ shows an initial value of 150, and after 8 hours, there are 54 left.

5. D Juan runs at a speed of 12 minutes per mile. The expression $12m$ represents the time it takes to run m miles. Adding in the 5 minutes warming up and the 10 minute cool down, the total time is $5 + 12m + 10$ minutes.

6.

Least		Greatest
Curt	Amber	Bailey

7. The total amount Marcella will spend on the new treadmill is __$873__.

8. B The expression that states how much was spent on apples and pears in total is $4 \times 2 + 3 \times p$, where p is the price per pound of pears. Setting this equal to $15.50 and simplifying gives the equation $8 + 3p = 15.5$. After subtracting 8 from both sides of the equation, we get the equation $3p = 7.5$, or $p = 2.5$ after dividing by 3.

Inequalities, p. 35

$<$	$\leq$	$>$	$\geq$
9. fewer than; less than; smaller than	**10.** less than or equal to; no more than; maximum	**11.** more than; larger than; greater than	**12.** no less than; greater than or equal to; minimum

13. A The inequality represented in this scenario is $300 + 7c \leq 1,000$, where c is the number of square feet of carpet. Subtracting 300 gives the inequality $7c \leq 700$, or $c \leq 100$ after dividing.

14. D Hannah's car can get up to 32 miles per gallon. Using m for miles, the inequality represented in this scenario is $\frac{m}{16} \leq 32$. Multiplying by 16 gives $m \leq 512$.

15. Maddie can spend __$55__ each week and still be able to buy the boots.

16. A Tara will spend $8p$ dollars on pizza and $15 on decorations. She can spend up to $60, her budget. Therefore, the inequality will be $8p + 15 \leq 60$.

17. C Jake will make $30x$ dollars for x hours. With his fee, the total will be $85 + 30x$, which must be greater than or equal to 500.

18. Using h to represent a person's height in inches, the inequality that represents the situation is __$h \geq 54$__.

19. The maximum amount Jolene can spend on shopping and still have enough for a $7 cab ride home is __$46__.

20. D The expression that represents Albert's commission earnings is represented by $0.1t$, where t is the total sales Albert makes a year. After adding in his salary, he wants to make more than 50,000. The phrase "more than" means the greater than symbol, $>$, is used.

Lesson 4.1

Identifying Polynomials, p. 37

1. C The polynomial $4x^2 + 6x - 3$ has three terms, which makes it a trinomial.

2. C The largest power of x in the polynomial is 2, which is its degree.

3. B A polynomial is in standard form when it is written in descending powers of x from left to right.

4. The value of the greatest exponent in the polynomial is the __degree__ of the polynomial.

5. After simplifying, the degree of the polynomial $(2x - 1)(3x^2 + 5)$ is __3__.

6. B Combine like terms and simplify. Simplifying other options gives a different coefficient for the x^2 or x^3 term.

Evaluating Polynomials, p. 38

7.

Least			Greatest
$x = 0; -1$	$x = 1; 0$	$x = -1; 4$	$x = -2; 15$

8. The value of the polynomial $5x^3 + x^2 - 10$ when $x = -3$ ___-136___.

9. **B** Substituting 2 into t in the polynomial $-16t^2 + 32t + 120$ gives the value $-16(2)^2 + 32(2) + 120 = -16 \times 4 + 64 + 120 = -64 + 64 + 120 = 120$ feet.

10. **D** The side length of the square is $3(5) - 2 = 13$ inches. The area is 169 square inches.

11. **A** Substituting 5 for m in the polynomial $-3m^2 + 39m - 48$ gives the value $-3(5)^2 + 39(5) - 48 = -3 \times 25 + 195 - 48 = -75 + 195 - 48 = 72$ pools.

Operations with Polynomials, p. 39

12. ___-1___$x^3 +$ ___1___$x^2 -$ ___4___$x -$ ___5___

13. **D** Multiply the two polynomials together making sure to distribute the terms in $(x-2)$ to every term in $(2x^2 - x + 3)$.

14. ▢ Select . . . ▼

 A. signs

15. **C** Distribute a negative one to each coefficient of the second polynomial. This will change the sign of each coefficient.

16. **B** The area of a rectangle is $A = lw$. Replacing the polynomials into the area formula, we get $(x-2)(x^2-1) = x^3 - 1x - 2x^2 + 2 = x^3 - 2x^2 - x + 2$.

17. ___$x^2 + x - 2$___ is the polynomial expression that represents the area of the triangle shown below.

18. **B** The perimeter of the rectangle is $P = 2(9x - 1) + 2(3x + 5) = 18x - 2 + 6x + 10 = 24x + 8$ feet.

Lesson 4.2

Factoring Out Monomials, p. 40

1. **D** The highest powers of x and y that can be factored out of each term are 1 and 2, respectively. The GCF of the coefficients is 15.

2. **D** When each term is divided by the GCF of the terms, the remaining portion is $2x^3y^2 + 3xy + 5$.

3. In the monomial, $11xy^2$, 11 is called the ___coefficient___.

4. **D** The GCF of the terms is $9ab$. When that is factored out, $b + 2a$ is left.

5. **C** The GCF of the terms is 16. When that is factored out, $x - 5$ is left.

6. ▢1 Select . . . ▼   ▢2 Select . . . ▼

 B. made an error by including x in the GCF   B. $3y(3x^2 - 2y + 4x)$

7. A ___monomial___ is a polynomial with one term.

8. **A** The GCF of the terms is 4. When that is factored out, $2mn^3 - 6m^2 - 3n^2$ is left.

9. **D** The GCF of the terms is y^2. When that is factored out, $2x - 5 - 10xy$ is left.

Factoring Quadratic Expressions, p. 41

10. ▢ Select . . . ▼

 B. second

11. The factored form of $x^2 - 5x - 24$ is $\underline{(x-8)(x+3)}$.

12. **C** Since the linear term is negative and the constant term is positive, both factors must be subtractions within each factor.

13. **B** The GCF of the terms is 4. When that is factored out, $x^2 - x - 6$ is left and factors to $(x-3)(x+2)$.

14. **D** The GCF of the terms is $2x$. When that is factored out, $6x^2 + x - 5$ is left, which factors to $(6x-5)(x+1)$.

15. **A** The GCF of the terms is 1, so nothing easily factors out. Take the product ac, and find factors that add up to 13. The product $ac = -30$ has factors 15 and –2 that add up to 13. Rewrite the polynomial as $6x^2 - 2x + 15x - 5 = 2x(3x-1) + 5(3x-1) = (2x+5)(3x-1)$.

16. **C** The GCF of the terms is x. When that is factored out, $8x^2 + 2x - 3$ is left. The product $ac = -24$, which has factors 6 and -4 that add up to 2. Rewrite the polynomial as $8x^2 - 4x + 6x - 3 = 4x(2x-1) + 3(2x-1) = (4x+3)(3x-1)$.

17. **B** The GCF of the terms is -16. When that is factored out, $t^2 + t - 6$ is left, which factors to $(t+3)(t-2)$.

18. $\underline{(x+6)}$ and $\underline{(x+7)}$ are two binomials that could represent the length and width of the rectangle.

Lesson 4.3

Solving a Quadratic Equation by Factoring, p. 43

1. **C** The polynomial factors as $(x-8)(x+7) = 0$. Setting both factors equal to 0 shows that the only positive solution is 8.

2. Using the zero-product principle, the solution to the equation $(x-4)(x+5) = 0$ is $\underline{x = 4 \text{ or } = -5}$.

3. The solution of the equation $x^2 - 19x = -90$ is $\underline{x = 10 \text{ or } x = 9}$.

4. **D** The product of x and $x + 1$ is represented by $x(x+1)$. This is equal to 72 and simplifies as $x^2 + x = 72$.

5. **A** The quadratic that models this scenario is $l(l-12) = 288$, or after simplifying $l^2 - 12l - 288 = 0$. This factors as $(l-24)(l+12) = 0$. Setting each term to 0 results in the positive solution $l = 24$ for the length, with 12 feet for the length.

Completing the Square, p. 44

6. **D** The coefficient of the linear term must be halved and squared to find c. $\frac{14}{2} = 7$ and $7^2 = 49$, which is what should be added to make the equation a perfect square.

7. The equation $x^2 - 8x + 16 = 0$ is a $\underline{\text{perfect square}}$ trinomial.

8. The solution that $x^2 - 32x = -256$ and $x^2 - 14x = 32$ have in common is $\underline{x = 16}$.

9. **B** Substitute $d = 36$ into the equation. This gives $36 = 16t^2$, or $t^2 = 2.25$. Taking square roots gives $t = \pm 1.5$, which only the positive value makes sense in the context.

10. **D** The equation $x^2 + 49 = 0$ has no real solutions. In order to solve it, you would need to take the square root of -49, which is not a real number. The other equations have real solutions.

11.
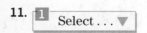 **1** Select . . . ▼

    C. subtract 39 from both sides of the equation

 **2** Select . . . ▼

    B. divide 16 by 2 to get 8

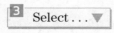 **3** Select . . . ▼

    A. add 64 to both sides of the equation

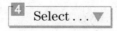 **4** Select . . . ▼

    D. $(x+8)^2 = 25$

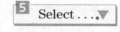 **5** Select . . . ▼

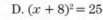   C. $-13, -3$

12. C The discriminant of the quadratic is $5^2 - 4(3)(10) = 25 + 120 = 145$. Since the discriminant is positive, there are two real solutions.

13. Select ... ▼

      B. The discriminant must be negative to have no real solutions, so $b^2 - 4(3)(12) = b^2 - 144 < 0$.
      This means that $b^2 < 144$, or $b < 12$.

14. A Using the quadratic formula, you can see that $a = 2$, $b = 5$, $c = -4$.

15. C Using the quadratic formula, you can see that $x = \dfrac{-8 \pm \sqrt{8^2 - 4(3)(-3)}}{2(3)} = \dfrac{-8 \pm \sqrt{64 + 36}}{6} = \dfrac{-8 \pm \sqrt{100}}{6} = \dfrac{-8 \pm 10}{6}$.

16. D Using the quadratic formula, you can see that $x = \dfrac{-16 \pm \sqrt{16^2 - 4(-16)(5)}}{2(-16)} = \dfrac{-16 \pm \sqrt{256 + 320}}{-32} = \dfrac{-16 \pm \sqrt{576}}{-32} = \dfrac{-16 \pm 24}{-32} = -0.25$ or 1.25. The negative answer does not relate to the situation, so $x = 1.25$.

Lesson 4.4

Simplifying Rational Expressions, p. 46

1. B A rational expression is an algebraic expression with a polynomial in the numerator and denominator. The other expressions do not have polynomials in the numerator or denominator.

2. B Substituting $x = -2$ into the expression results in
$$\dfrac{2(-2)^2 + 3(-2) - 2}{-(-2) + 4} = \dfrac{2(4) - 6 - 2}{2 + 4} = \dfrac{8 - 6 - 2}{6} = \dfrac{0}{6} = 0.$$

3. C The restricted values are the values that make the denominator equal to 0. The denominator can be factored as $x^3 - 9x = x(x - 3)(x + 3)$.

4. A The numerator can be factored as $x^3 - 4x = x(x^2 - 4)$, which has the denominator as a factor and can be canceled out. The expression x is what remains along with the restricted values from $x^2 - 4$.

5. Select ... ▼

      D. undefined

6. The _restricted values_ for the rational expression $\dfrac{x^2 + 4x}{x^2 - 16}$ are -4 and 4 because a rational expression is undefined when the denominator is equal to 0.

Multiplying and Dividing Rational Expressions

7. C Multiplying rational expressions does not require that each rational expression have the same denominators. Therefore, simplifying first will help with the multiplication.

8. B The denominators are $2x^2 + 8x$ and $x^2 - 3x$. These can be factored as $2x(x + 4)$ and $x(x - 3)$. The restricted values can be found from the factored denominators.

9. The expression simplified is $\dfrac{x^2 + 2x - 8}{2x^2}$.

10. D Plugging in -3 for x gives the value $\dfrac{(-3)^2 + 2(-3) - 8}{2(-3)^2} = \dfrac{9 - 6 - 8}{2(9)} = \dfrac{-5}{18}$. The value 3 is one of the restricted values and cannot be evaluated in the rational expression.

11. B The restrictions need to be found for both denominators as well as the numerator $x^2 + 4x + 3$ since division by 0 is undefined. The denominators factor as $x^2 + x - 12 = (x - 3)(x + 4)$ and $x^2 - 4 = (x - 2)(x + 2)$, while the numerator factors as $x^2 + 4x + 3 = (x + 1)(x + 3)$.

12. Therefore, the expression that describes the length of the pool is _$x + 3$_.

Adding and Subtracting Rational Expressions, p. 48

14. **D** The common denominator for the two rational expressions is $(x-6)(x+2) = x^2 - 4x - 12$. The numerator will become $(x+3)(x+2) + (x-4)(x-6) = 2x^2 - 5x + 30$.

15. **C** Substituting 0 into the expression simplifies as $\dfrac{2(0)+3}{0^2-16} - \dfrac{4(0)-4}{(0)^2-1} = \dfrac{3}{-16} - \dfrac{-4}{-1} = -\dfrac{3}{16} - 4 = -\dfrac{67}{16}$. The value 4 is a restricted value and is undefined.

16.
Select . . . ▼

B. $\dfrac{2x+2}{x^2+2x}$

17.
Select . . . ▼

C. $\dfrac{2x-1}{x^2-x-6}$

18. The expression to describe how long it will take Marge to complete the loop on her bike is __x + 1__.

19. If the Rapids is twice as fast as the Bayou, then the expression representing how close they will be to crossing each other's path in an hour is ___$\dfrac{3}{2x}$___. If it takes the Rapids 2 hours to complete the trip, then it will take them _about 1.3_ hours to cross each other's path.

Lesson 5.1

Points and Lines in the Coordinate Plane, p. 49

1.

Quadrant I	Quadrant II	Quadrant III	Quadrant IV
B	A	E	C, D

2. **C** The graph shows the equation of the line $y = -1$. Therefore, any coordinate with -1 as its y-coordinate will be a solution to the equation.

3. (Sample Answer)

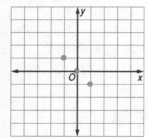

4. Table ____D____ represents the equation $y = -x + 2$.

The Slope of a Line, p. 50

5. **B** Using the coordinates $(-2, -4)$ and $(-1, -2)$, you can find the slope by finding the change in y divided by the same change in x, or $\dfrac{-4-(-2)}{-2-(-1)} = 2$

6. **B** To approximate the slope of a line, look at the steepness of it. Graphs A and C have negative slope, and Graph D has a slope greater than 1 because it travels more than 1 up going one unit across.

7.

Least Expensive			Most Expensive
Web site D	Web site A	Web site B	Web site C

Slope as a Unit Rate, p. 51

8. D   The hourly rate for Painter A is \$25. The hourly rate for Painter B is \$35. Therefore, Painter A charges \$10 less an hour, not \$5 less an hour, than Painter B.

9. After two and a half hours, rower _____B_____ travels farther by ____12.5____ miles.

10. C   The unit rate is the number of words texted per minute. Finding the slope of the line gives 50 words per minute.

11. D   The unit rate is the number of dollars per item. Finding the slope of the line gives \$3 per item.

Lesson 5.2

Using Slope and y-Intercept, p. 52

1. A   Subtracting $0.25x$ from both sides of the equation gives $-0.25x + y = -5$. Since the coefficient of x must be a whole number, multiplying the entire equation by -4 will rewrite the equation in standard form, or $x - 4y = 20$.

2. D   The unit rate of 3 translates to the slope of 3, and the initial value of 2 means that the y-intercept is 2. Therefore, the equation of the line is $y = 3x + 2$.

3. B   Using point-slope form, the equation of the line is $y - 4 = 3(x - 2)$. The equation can be simplified to $y = 3x - 2$.

4. C   The line has a slope of 1 and passes through the point $(3, 8)$. Therefore, the equation of the line is $y - 8 = 1(x - 3)$, or $y = x + 5$.

5. B   The line has a slope of 4 and has a y-intercept of 8. Therefore, the line is $y = 4x + 8$.

Using Two Distinct Points, p. 53

6. D   The slope of the line between the two points is $\frac{-2 - (-1)}{-3 - (-5)} = -\frac{1}{2}$.

7. B   The intercepts of the line are $(0, 3)$ and $(2, 0)$. The slope of the line is $-\frac{3}{2}$ and the equation of the line is $y = -\frac{3}{2}x - 2$. Multiplying by 2 to both sides simplifies to $2y = -3x + 6$, which is the same as $3x + 2y = 6$.

8. A   The slope of the line is -2 and so the equation of the line is $y - 2 = -2(x + 1)$, or $y = -2x$.

9.

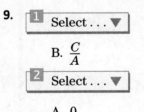

B. $\frac{C}{A}$

   2 Select . . . ▼

A. 0

3 Select . . . ▼

A. 0

4 Select . . . ▼

B. $\frac{C}{B}$

10. Standard Form __3__ $x +$ __4__ $y =$ __−12__

Slope-Intercept Form $y =$ __$-\frac{3}{4}$__ $x +$ __−3__

Point-Slope Form $y -$ __−3__ $= m(x -$ __0__$)$ OR $y -$ __0__ $= m(x -$ __−4__$)$

Using Tables, p. 54

11.

x	y
−3	−6
−2	−4
−1	−2
0	0
1	2
2	4
3	6

12. B The equation of the line that represents the data is $y = 9x + 7$. The other answer choices correctly align to the data/equation.

13.

x	y
0	4
3	2

14. ___ $y = 3x$ ___ is the equation of the line represented by the table.

15.

x	y
−2	−7
−1	−4
0	−1
1	2
2	5
3	8
4	11

16.

x	y
−4	−1
−2	$-\dfrac{1}{2}$
0	0
2	$\dfrac{1}{2}$
4	1

Lesson 5.3

Using Ordered Pairs, p. 55

1. B Solving the equation for y in the equation gives $y = 12x - 4$. When $x = 0$, then $y = -4$ by substituting 0 into the equation. When $x = 1$, then $y = 12(1) - 4 = 8$.

2.

x	y
−2	1
−4	5
4	−11

3. D Graph B and D pass through the point (0, 1). Only Graph D passes through (1, 4).

Using Slope-Intercept Form, p. 56

4. **C** A line with negative slope moves from the top left to the bottom right of the graph. A line with a positive y-intercept crosses the y-axis above the origin.

5. **B** The slope of the line is 2 units down and 1 unit left. Both down and left are negative directions and so the slope would be $\frac{-2}{-1} = 2$. Harold should have moved right instead of left.

6.
1 Select . . . ▼		2 Select . . . ▼

  B. y-intercept   A. slope

7. **C** Since Ted is returning the shirts, he is being reimbursed for each shirt and is therefore paying less in total.

8.
Select . . . ▼

  B. Layla

9.
Select . . . ▼

  B. become less steep

10. According to the graph pictured below, Matt joined a music sharing program that costs $ _____10_____ initially and $ _____2_____ for every song he downloads.

11.

Number of Laps

12. **B** All other choices mention scenarios with positive slope. Taking money out of an account represents a negative slope.

Lesson 5.4

The Graphing Method, p. 58

1. **C** Solving the two equations gives the solution $(-2, 10)$.

2. Using the graphing method to solve, you would graph the equation _____$y = 1.25x + 25$_____ for Tow-A-Way and the equation _____$y = 1.5x + 15$_____ for Haul-Ur-Car, and then determine the solution from the graph. The solution is $(40, 75)$.

The Substitution Method, p. 59

3.
Select . . . ▼

  B. no solution

4. **D** Solving the first equation for x and substituting into the second equation gives the solution $x = 1$. Plugging this value for x into either equation and solving for y gives $y = 0$.

5. **B** The two equations are $y = 4.5x + 100$ and $y = 12.5x + 20$. Setting them equal to each other gives the value $x = 8$.

6. **D** The two equations that are to be solved are $y = 1.15x + 25.95$ and $y = 1.25x + 19.95$. Setting them equal to each other gives the value $x = 60$.

7. Assuming that the trend of increasing and decreasing continues, the equation $\underline{y = -50x + 12{,}500}$ would represent Water for You and the equation $\underline{y = 40x + 8{,}000}$ would represent Drink Up. After solving this system of equations, it is found that it will take $\underline{\hspace{2em}50\hspace{2em}}$ months for each company to sell $\underline{\hspace{1em}10{,}000\hspace{1em}}$ containers a month.

The Elimination Method, p. 60

8.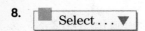

   C. infinitely many solutions

9. **A** Multiply the first equation by 4, the second equation by -3, then add the two equations together. This will result in $11y = -88$, or $y = -8$. Substitute the value into either equation. Simplifying will result in $x = -8$.

10. **A** The equations needed to solve are $s + a = 500$ and $5s + 8a = 3{,}700$, where s and a represent the number of student tickets and adult tickets sold, respectively. Multiplying the first equation by 5 and subtracting it from the second equation gives the equation $3a = 1{,}200$, or $a = 400$. Substituting that value into either equation, you can find that $s = 100$.

11. **C** The equations needed to solve are $5j + 9t = 376$ and $2j + 6t = 208$, where j and t represent the cost of jeans and T-shirts, respectively. Multiplying the first equation by 2, the second equation by -5, then adding them together results in the equation $-12t = -288$, or $t = 24$. Substituting that value into either equation, you can find that $j = 32$.

12. **B** The equations needed to solve are $x + y = \dfrac{1{,}200}{5}$ and $x - y = \dfrac{1{,}200}{6}$ where x and y are the speed of the plane and wind in miles per hour, respectively. Adding the two equations together gives $2x = 440$, or $x = 220$. Substituting that value into either equation, you can find that $y = 20$.

Lesson 6.1

Functions, p. 61

1. **B** Horizontal lines can be drawn through multiple points on graphs A, C, and D.

2. **D** Substituting 1 into the function gives $f(1) = 0.49(1) + 44.95 = 45.44$.

3. $\underline{\$167.45}$ is the total cost of printing 250 flyers.

Linear and Quadratic Functions, p. 62

4.

Least			Greatest
$f(3)$	$g(0)$	$g(-3)$	$f(0)$

5. **A** The values for the function f are as follows: $f(-1) = 8; f(0) = 1; f(1) = 0$.

6. **D** Multiply the number of gallons by 3.18.

7. **B** The function is linear because the only power of x is 1. All non-horizontal linear functions are one-to-one.

8. **A** Substituting the values into the function gives $f(2.5) = -4.9(2.5)^2 + 19.6(2.5) + 98 = -4.9(6.25) + 49 + 98 = -30.625 + 147 = 116.375$ and $f(1.5) = -4.9(1.5)^2 + 19.6(1.5) + 98 = -4.9(2.25) + 29.4 + 98 = -11.025 + 127.4 = 116.375$. Therefore, the difference is 0.

Functions in the Coordinate Plane, p. 63

9.

x	f(x)
−3	−18
−1	8
0	4
3	2

10.

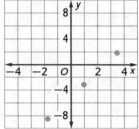

11. **B** The *x*-value that splits the graph is 1. The left part of the graph represented by $y = 4x + 5$ passes through the *y*-axis at (0, 5). The right part of the graph does not pass through the *y*-axis, but its left-most point is (1, 1).

12.

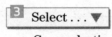

Lesson 6.2

Evaluating Linear and Quadratic Functions, p. 64

1. **A** The common difference of a linear equation is its slope.

2.

[1] Select . . . ▼	[2] Select . . . ▼	[3] Select . . . ▼
C. 4	B. 2nd	C. quadratic

3. **C** Plugging in −2 into the function gives the value −25. Since the slope is 11, for every increase of 1 in the *x*-value, there will be an increase of 11 in the *y*-value.

4. **A** Plugging in 0 into the function gives the value 3. Plugging in 1 into the function gives the value 2.

5. **B** Plugging in 0 into the function gives the value −8. Plugging in 1 into the function gives the value −5.

Recognizing Linear and Quadratic Functions, p. 65

6.

x	f(x)	1st Cons. Diff.	2nd Cons. Diff.
−2	−19	10	0
−1	−9	10	0
0	1	10	0
1	11	10	0
2	21	10	
3	31		

The table represents a ___linear function___.

7. The common difference is ___0___.

x	f(x)	1st Cons. Diff.
−2	−2	0
−1	−2	0
0	−2	0
1	−2	0
2	−2	0
3	−2	

8. **C** A quadratic function has a 2nd common consecutive difference.

9.

Least		Greatest
h(x)	g(x)	f(x)

10. **D** The last column is a common difference. Since it is the 4th consecutive difference, the polynomial is a 4th degree polynomial.

x	f(x)	1st Cons. Diff.	2nd Cons. Diff.	3rd Cons. Diff.	4th Cons. Diff.
−3	83	−65	50	−36	24
−2	18	−15	14	−12	24
−1	3	−1	2	12	24
0	2	1	14	36	
1	3	15	50		
2	18	65			
3	83				

11. Helen is _____incorrect_____.

Lesson 6.3

Key Features, p. 67

1. A. line of symmetry [$x = 1$]

B. relative minimum [none]

C. relative maximum [16]

D. x-intercept(s) [−3, 5]

E. y-intercept(s) [15]

2. **B** Graph C has a y-intercept of −5, while graphs A and D are quadratic functions.

3. **C** The graph falls from the left until $x = -1$, rises until $x = 5$, then falls again.

4. **A** The graph rises to the left and falls to the right.

5. **C** The largest y-value that the graph reaches is when $x = -5$ at 6.

6. **C** The smallest y-value that the graph reaches is when $x = 3$ at −10.

7. A. The graph shows ___rotational___ symmetry.

B. The end behavior on the left side extends indefinitely _____up_____.

C. The end behavior on the right side extends indefinitely _____down_____.

D. The decreasing interval is __all the values of y__.

Use Key Features to Draw a Graph, p. 69

8.

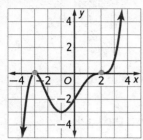

9.

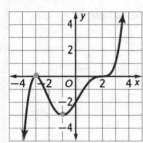

10. C   The left side of the graph extends down while right side of the graph extends up.

11. D   Graphs A and B do not extend down indefinitely, while graph C does not have an x-intercept at 2.

Lesson 6.4

Compare Proportional Relationships, p. 70

1. B   The Get Fit Gym charges $45 for 3 months, which is more than what the Move More Gym charges.

2. C   E-Your Book pays $1 for 20 books, or 5 cents per book, which is not the same as Sell Your Book.

Compare Linear Functions, p. 71

3. B   The cost of the 13WSR bulb is $7 with a $1.50 per month operating cost.

4. According to the data for Fresh Day and Friendly Earth, Fresh Day will become more expensive than Friendly Earth when greater than _____1_____ pound(s) of almonds is sold.

5. C   For 3 hours, the plumber charges $165 while the electrician charges $150.

Compare Quadratic Functions, p. 72

6. Mr. Bott's golf ball will hit the ground after _____8_____ seconds and Mrs. Bott's golf ball will hit the ground after _____3_____ seconds.

7. Mrs. Bott's golf ball will reach a maximum height of _____64_____ feet.

8.
┌─────────────────────┐
│ Select . . . ▼   │
└─────────────────────┘
   B. older

9. The older book will start making a profit after _____2_____ weeks and stop making a profit after _____15_____ weeks while the newer book will start making a profit after _____9_____ weeks and stop after _____20_____ weeks.

Lesson 7.1

Rectangles, p. 73

1. D   The area of the driveway is $35 \times 24 = 840$ square feet. With two coats, that will be 1,680 square feet. Since each tub coats 450 square feet, you will need $\frac{1,680}{450} = 3.7$, or 4 tubs. For $18.50 per tub, you will spend $74.

2. 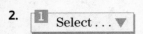   

     B. perimeter   C. twice

3.   __0.15 m²__   is the area of the rectangle.

4.   C   The perimeter of the school is $P = 2(80) + 2(52) = 264$ feet. The number of laps needed to run 1 mile is $\frac{5,280}{264} = 20$ laps.

5.   D   The ribbon is being wrapped up the front (10 inches), across the top (12 inches), down the back (10 inches), and the bottom (12 inches). The total of that is 44 inches.

6.   B   The length of the cutting board will be the area divided by the width, or $\frac{1,140}{30} = 38$ centimeters.

Triangles, p. 74

7.   C   Using the Pythagorean Theorem, the hypotenuse is $h = \sqrt{8^2 + 15^2} = \sqrt{64 + 225} = \sqrt{289} = 17$ yards. The perimeter will be the sum of the sides, or 40 yards.

8.   C   The area of the triangle is $7.5 = \frac{1}{2}(3)(h)$. Multiplying by 2 and dividing by 3 gives $5 = h$.

9.   The area of the following triangle is __30 mm²__.

10. C   The third side of the triangle is either 8 inches or 3 inches. If it is 3 inches, then the third side won't connect to the side of length 8. Therefore, the missing side must be 8 inches long and the perimeter will be 19 inches.

11. Length of side $b =$ | 20 ft |

     Perimeter = | 60 ft |

        Area = | 150 ft² |

12.   __11__   is the length of side a.

13.

         Return: _6_ : _00_   *Start:*12:00

      B: _3_ : _30_   *A:* _1_ : _30_

Parallelograms and Trapezoids, p. 75

14. B   The height of the parallelogram can be found by using the Pythagorean Theorem, $h = \sqrt{25^2 - 7^2} = \sqrt{625 - 49} = \sqrt{576} = 24$ ft. The area is $24 \times 7 = 168$ ft².

15. B   The height of the parallelogram can be found by using the Pythagorean Theorem, $h = \sqrt{15^2 - 9^2} = \sqrt{225 - 81} = \sqrt{144} = 12$ in. The area is $12 \times 9 = 108$ in².

16. D   The area of the trapezoid is $35 = \frac{1}{2}(3 + b)5$. Multiplying by 2 and dividing by 5 results in $14 = 3 + b$, or $b = 11$ meters.

17. A   The area of the parallelogram can be found in different ways, but both are equal to each other. $8 \times h = 6 \times 4$, or $8h = 24$. This simplifies to $h = 3$ mm.

18. What is the height of the trapezoid?   __8 cm__

Lesson 7.2

Circumference, p. 76

1.   C   The circumference of the circle is $600 \times 3.14 = 1,884$ feet. Traveling at 150 ft/sec, it will take $\frac{1,884}{150} = 12.56$, or 12.6 seconds.

2. B   The distance from the Entrance to the Fountain is half of the circumference, or $\frac{1}{2} \times 800 \times 3.14 = 1{,}256$. The difference is $1{,}256 - 800 = 456$.

3. C   The circumference of the bicycle is $0.67 \times 3.14 = 2.1038$. The number of times the wheel must revolve is $\frac{1000}{2.1038} = 475.3$, or 475 times.

4. B   When the large gear rotates 15 times, a total distance of 150π is traversed. For the small gear to travel the same distance, it will have to turn $\frac{150\pi}{6\pi} = 25$ times.

5. D   After 15 seconds, the circle will have a radius of 30 feet, or a circumference of 188.4 feet.

6.   If the blades spin at the rate of 300 revolutions per minute, then to the nearest 10 feet, the tips of the blades will travel _____8,480_____ feet in 30 seconds (to the nearest 10 feet), using 3.14 for π.

Area, p. 77

7.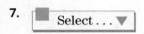

    C. radius

8. B   The radius of the circle is $\frac{22\pi}{2\pi} = 11$, and so the area is 121π.

9.   To the nearest tenth of an inch, the minimum diameter her new table must have is _____21.2_____ inches, using 3.14 for π.

10.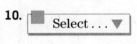

    B. 27

11. B   The shape is composed of a square and two semi-circles. The total area will be $A = 4^2 + 3.14 \times 2^2 = 16 + 4 \times 3.14 = 28.56$, or 28.6 cm^2.

12. B   The area of the large circle is $10.5^2 \times 3.14 = 346.185$. The area of the inside circle is $7^2 \times 3.14 = 153.86$. The shaded part is the difference of the two areas, or 192.325 or 192 m^2.

13.   The area of a circle that has a circumference of 18π is _____81_____ π.

14.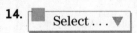

    B. Raffi

Find Radius or Diameter, p. 78

15. A   The circumference of the track is $\frac{5{,}280}{4} = 1{,}320$ feet. Setting it equal to the formula gives $1{,}320 = 2 \times 3.14 \times r$, or $r = \frac{1{,}320}{6.28} = 210.2$, or 210 feet.

16.   To the nearest foot, the radius of the largest circular lawn you can plant with the grass seed that you have is _____16_____ feet when you use 3.14 for π.

17. A   The diameter of the ferris wheel is $\frac{126}{3.14} = 40.1$, or 40 feet.

18. D   The ratio of diameter to cups of flour is 8 to 2, or 4 to 1. Multiply $4\frac{1}{2}$ by the ratio to get the diameter, or 18 in.

Lesson 7.3

Rectangular Prisms, p. 79

1. B   Volume is a 3-dimensional measurement and surface area is a two-dimensional measurement.

2. C   The volume can be found by multiplying the 3 dimensions, or $8.5 \times 9.5 \times 14 = 1{,}130.5$ in.^{3}

3. D   The surface area can be found by finding the area of each side of the prism and adding them together, or $2(8.5 \times 9.5) + 2(8.5 \times 14) + 2(9.5 \times 14) = 665.5$ in.^{2}

4. **A** The volume can be found by multiplying the 3 dimensions, or $50 \times 16 \times 1 = 800$ ft.^{3}

5. **B** The visible area is the top and sides, which have surface area
$50 \times 16 + 2(50 \times 1) + 2(16 \times 1) = 800 + 100 + 32 = 932$ ft.^{2}

6. The height of the container is ___18mm___.

Cylinders and Prims, p. 80

7. **B** The volume can be found by applying the formula for the volume of a cylinder, or
$V = 3.14(4)^2 (20) = 3.14 \times 16 \times 20 = 1{,}004.8$ cm.^{3}

8. **B** The can has 3 surfaces: the top, bottom, and side. The area of each part is $3.14(4)^2$, $3.14(4)^2$, and
$3.14 \times 8 \times 20$. The total surface area is $2(3.14(4)^2) + 3.14 \times 8 \times 20 = 602.88$, or 602.9 in.^{2}

9. ___1,200 in.^{3}___ is the volume of the prism.

10. The surface area is ___920 in.^{2}___

11. **C** The volume of the prism is the area of the base times the height, or
$V = \left(\frac{1}{2} (2)(3) \right) \times 4 = 3 \times 4 = 12$ ft.^{3}

12. **B** The surface area can be found by finding the area of the 5 sides, or
$2\left(\frac{1}{2} (2)(3) \right) + 2(4 \times 2.5) + 3 \times 4 = 6 + 20 + 12 = 38$ ft.^{2}

Pyramids, Cones, and Spheres, p. 81

13. **D** The volume of the pyramid is $V = \frac{1}{3}(12 \times 12)(8) = 384$ cm^3.

14. **B** The slant height of the pyramid can be found by using the Pythagorean Theorem, or
$s = \sqrt{6^2 + 8^2} = \sqrt{36 + 64} = \sqrt{100} = 10$ centimeters. The surface area of the square pyramid is
$(12 \times 12) + 4\left(\frac{1}{2} (12)(10) \right) = 144 + 240 = 384$ cm.^{2}

15. **A** The volume of the cone is $V = \frac{1}{3}(3.14(5)^2) \times 12 = 314$ in.^{3}

16. **C** The slant height of the cone can be found using the Pythagorean Theorem, or
$s = \sqrt{5^2 + 12^2} = \sqrt{25 + 144} = \sqrt{169} = 13$ inches. The surface area of the cone is
$3.14(5)(13) + 3.14(5)^2 = 204.1 + 78.5 = 282.6$ in.^{2}

17.

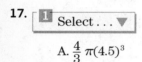

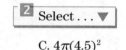

A. $\frac{4}{3} \pi(4.5)^3$

C. $4\pi(4.5)^2$

18. The height, to the nearest foot, is ___23 feet___.

Lesson 7.4

2-Dimensional Figures, p. 82

1. **D** The perimeter can be found by adding up the lengths of all 16 sides, or $2 \times 12 + 14 \times 3 = 66$ inches.

2. **C** The area can be found by adding up the areas of the 4 pieces, or $4(3 \times 12) = 144$ square inches.

Volume of 3-Dimensional Solids, p. 83

3. **C** The solid is composed of two figures, a rectangular prism and a square pyramid. The volume of
each is $(8)(8)(5)$ and $\frac{1}{3}(8)(8)(3)$, respectively. The total volume is $320 + 64 = 384$ mm^3.

4. **B** The side length of the cube is $2\frac{1}{2}(8) = 20$ centimeters. The volume of the cube is 8,000 cm^3, while
the volume of the pyramid is $\frac{1}{3}(20)(20)(8) = 1{,}066.6$. The total volume is 9,066.6, or 9,067 cm^3.

5. **B** The height of the cylinder in the middle is 6mm, and the height, or radius, of each hemisphere is 2.
The volume of the capsule is $\frac{4}{3} \pi(2)^3 + \pi(2)^2 (6) = 108.9$, or 109 mm^3.

6. The farmer can store ___38,858___ cubic feet of grain in the silo.

7. 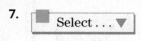 Select . . . ▼

C. 2,199 cm³.

Surface Area of 3-Dimensional Solids, p. 84

8. **B** The figure has 9 faces. The faces of the prism are $(4)(5)(8)$ and $(8)(8) = 224$. The faces of the pyramid are $(4)(\frac{1}{2})(8)(5) = 80$. The total surface area is the sum of the faces of the pyramid and the faces of the prism, 304 mm².

9. If a gallon of paint covers 400 square feet, the farmer needs ___15___ gallons of paint.

10. 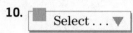 Select . . . ▼

A. 126 mm².

11. The surface area of the perfume bottle box is ___1,036 cm²___.

12. **D** The volume of the basketball is given, and you need to find the surface area. The radius of the sphere can be found using the equation $121.5\pi = \frac{4}{3}\pi r^3$, or $r^3 = 91.125$. This gives $r = 4.5$. The surface area of a sphere can be found using $S = 4\pi(4.5)^2 = 4\pi(20.25) = 81\pi$.

13.

Least Surface Area			Greatest Surface Area
Square pyramid with base length 5 and slant height 6	Cube with side length 5	Sphere with radius 5	Cylinder with radius 5 and height 6

Lesson 8.1

Measures of Central Tendency, p. 85

1. **B** The sum of the data is $32 + 31 + 33 + 44 + 35 + 44 + 37 + 44 + 40 + 41 + 38 + 38 + 40 + 32 + 44 = 573$. With 15 data points, the mean is $\frac{573}{15} = 38.2$.

2. **A** The value 44 occurs 4 times within the data set.

3. **C** In order from least to greatest, the set is 31, 32, 32, 33, 35, 37, 38, 38, 40, 40, 41, 44, 44, 44, and 44. The median is the 8th value from the left.

4. 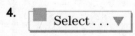 Select . . . ▼

D. range

5. 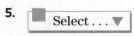 Select . . . ▼

A. mean

Finding a Missing Data Item, p. 86

6. **D** The average can be found by solving the equation $\frac{78 + 82 + 86 + x}{4} = 84$. Multiply by 4 and simplify both sides, $246 + x = 336$. Subtract 246 from both sides gets $x = 90$.

7. **C** To find the sales required, solve the equation $\frac{900 + 1,500 + 785 + 895 + 973 + 1,100 + 875 + x}{8} = 1,000$. Multiply by 8 and simplify to get $7,028 + x = 8,000$. Subtract 7,028 to get $x = 972$.

8. **B** There are a total of 4 tests that need to average to a 90% for Allison to get the grade she wants. That equation would be $\frac{87 + 85 + 92 + x}{4} = 90$.

9. Can Wes reach his goal if the next test doesn't have any extra credit on it? ___No___

10. Is it possible for Karina to meet her goal? ___No___

Weighted Averages, p. 87

11. The mean number of jeans owned by those surveyed is _____3_____.

12. ___0 and 1___ are the number of jeans they would each need to own so that the average number of pairs of jeans owned is 2.75.

13. B The total number of sessions is $25 + 30 + 15 = 70$. The weighted average is
$$\frac{75(30) + 150(30) + 200(15)}{70} = 133.93.$$

14. A The total number of baked goods is $30 + 12 + 40 = 82$. The weighted average is
$$\frac{30(3) + 12(20) + 40(4)}{82} = 5.98.$$

15. C Since the final is weighted as 3 tests, the total number of tests is 6. The weighted average is
$$\frac{87 + 82 + 94 + 3(91)}{6} = 89.3.$$

Lesson 8.2

Bar Graphs, pp. 88–89

1. D The sales for the 1st 3 quarters are $40,000, $100,000, and $80,000. The equation to find the 4th quarter sales is $\frac{40,000 + 100,000 + 80,000 + x}{4} = 70,000$. Multiplying by 4 and simplifying gives $220,000 + x = 280,000$, or $x = 60,000$.

2. C The percent increase is $\frac{100,000 - 40,000}{40,000} = \frac{60,000}{40,000} = 1.5$ or 150%.

3. B The interest for the 30-year loan is about $270,000, while the interest for the 15-year loan is about $115,000. The difference is $155,000.

4. D The 30-year interest on a 9% loan is about $190,000.

5. Over the life of a 30-year mortgage at 6%, ___$4,000___ is the average interest paid per year for a mortgage of $100,000.

6. B The other choices give either too many or not enough ticks on the bar graph.

Circle Graphs, p. 90

7. Gregg received _____6_____ percent of the votes.

8. ___Pullam___ received about $\frac{1}{5}$ of the votes cast.

9. D The total number of registered voters is not given.

10. D Lee received 25% of the 50,200 votes cast, or $0.25 \times 50,200 = 12,550$.

11. 5% ___bicycle___ 20% ___walk___ 30% ___bus___ 45% ___car___

12. A The estimated profits for each quarter are $3,000, $3,500, $4,000, and $7,500. The average of those numbers is $\frac{3,000 + 3,500 + 4,000 + 7,500}{4} = 4,500$.

13. D The total profit is $18,000. The first quarter made 16.7% of that, or 0.167 of it.

Lesson 8.3

Dot Plots, p. 91

1. B There number of people Cassie surveyed are $2 + 4 + 3 + 2 = 11$.

2. B The 5th number from the left is the median, 1.

3. C Of those surveyed, 5 own at least 2 computers. This fraction is $\frac{5}{11}$.

4. The data set for this dot plot, from least to greatest is ___0, 0, 1, 1, 1, 1, 2, 2, 2, 3, 3___.

5.

Least	3 computers	Greatest
0 computers		1 computer

Histograms, p. 92

6. **D** The numbers in each category are 3, 9, 15, 6, 4, 2, and 1, for a total of 40.

7. **B** The intervals are equally spaced and so the range is the same for each interval. The first interval is 15–19, which covers the years 15, 16, 17, 18, and 19, a total of 5 years.

8. **C** The number in the right 3 categories are 4, 2, and 1, a total of 7.

9. **D** There are 27 first-time mothers younger than 34 in the data, or $\frac{27}{40} = 67.5\%$.

10. A possible data set for this histogram, in order from least to greatest, is <u>19, 19, 19, 20, 20, 21, 21, 21, 22, 22, 23, 23, 25, 25, 25, 26, 26, 26, 27, 27, 27, 28, 28, 28, 29, 29, 29, 30, 30, 31, 32, 33, 34, 35, 36, 37, 38, 40, 41, 45</u> .

Box Plots, p. 93

11. For the box plot above, the median is <u>14</u> , 1st quartile is <u>12</u> , and the 3rd quartile is <u>15</u> .

12. A possible data set for this box plot is <u>10, 11, 12, 13, 13, 14, 14, 15, 15, 16, 17</u> .

13. **C** The median of the data is the middle point, or 14.

14. **A** The range is the greatest data point minus the least data point, or 17 − 10 = 7.

15. **D** The numbers between the points on the box plot represent exactly one-half of the data. The other statements are not true for the data shown.

16.

1 Select . . . ▼

A. 9

2 Select . . . ▼

C. 45

3 Select . . . ▼

C. 37.5

4 Select . . . ▼

C. median, 25

17. **B** The whiskers from 4–8 and 16–20 represent half of the data. Since there are 3 values in both the 4–8 and 16–20 whiskers, placing another value in one whisker would cause a miscalculation in the other whiskers. The other options either lie in the whiskers or are outliers.

Lesson 8.4

Tables, p. 94

1.

	1 side	2 sides	3 sides
Vegetables	$1.29	$2.39	$3.19
Potatoes	$1.79	$2.99	$3.89
Fruits	$1.49	$2.89	$4.19

2. <u>5</u> columns, <u>4</u> rows OR <u>4</u> columns, <u>5</u> rows

3. **D** During week 3, there were 472 bacteria and 108 birds, which is 364 more.

4. **C** In weeks 3 and 4, there were 472 and 981 bacteria, respectively. In that same time, there were 268 + 108 = 376 and 319 + 102 = 421 ants and birds in weeks 3 and 4 respectively.

Scatter Plots, p. 95

5.

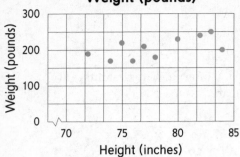

Weight (pounds)

6.

Dollars Spent

Line Graphs, p. 96

7. D The other options would be more appropriate to display on a scatter plot.

8.

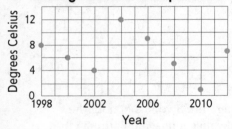

Average Winter Temperature

9. D The graph shows no trend and cannot be described by the other choices.

10. C The other choices add up to too little or too much.

11.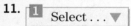

B. 2004

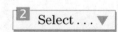

D. 2010

Area of a	square	Area = side^2
	rectangle	Area = length × width
	triangle	Area = $\frac{1}{2}$ × base × height
	parallelogram	Area = base × height
	trapezoid	Area = $\frac{1}{2}$ × (base$_1$ + base$_2$) × height
	circle	Area = π × radius^2; π is approximately equal to 3.14
Perimeter of a	square	Perimeter = 4 × side
	rectangle	Perimeter = 2 × length + 2 × width
	triangle	Perimeter = side$_1$ + side$_2$ + side$_3$
Circumference of a	circle	Circumference = π × diameter; π is approximately equal to 3.14
Surface Area of a	rectangular/ right prism	Surface Area = 2(length × width) + 2(width × height) + 2(length × height)
	cube	6 × side^2
	square pyramid	Surface Area = ($\frac{1}{2}$ × perimeter of base × height of slant) + (base edge)^2
	cylinder	Surface Area = (2 × π × radius × height) + (2 × π × radius^2); π is approximately equal to 3.14
	cone	Surface Area = (π × radius × height of slant) + (π × radius^2)
	sphere	Surface Area = 4 × π × radius^2
Volume of a	rectangular/ right prism	Volume = length × width × height
	cube	Volume = edge^3
	square pyramid	Volume = $\frac{1}{3}$ × (base edge)^2 × height
	cylinder	Volume = π × radius^2 × height; π is approximately equal to 3.14
	cone	Volume = $\frac{1}{3}$ × π × radius^2 × height
	sphere	Volume = $\frac{4}{3}$ × π × radius^3

Coordinate Geometry	(x_1, y_1) and (x_2, y_2) are two points in a plane.
	slope of a line $= \dfrac{y_2 - y_1}{x_2 - x_1}$; (x_1, y_1) and (x_2, y_2) are two points on the line
	slope-intercept form of the equation of a line $y = mx + b$, when m is the slope of the line and b is the y-intercept
	point-slope form of the equation of a line $y - y_1 = m\,(x - x_1)$, when m is the slope of the line
Pythagorean Relationship	$a^2 + b^2 = c^2$; in a right triangle, a and b are legs, and c is the hypotenuse
Quadratic Equations	standard form of a quadratic equation $ax^2 + bx + c = 0$
	quadratic formula $x = \dfrac{-b \pm \sqrt{b^2 - 4ac}}{2a}$
Measures of Central Tendency	$\textbf{mean} = \dfrac{x_1 + x_2 + \ldots + x_n}{n}$, where the x's are the values for which a mean is desired, and n is the total number of values for x
	$\textbf{median} =$ the middle value of an odd number of ordered scores, and halfway between the two middle values of an even number of ordered scores
Simple Interest	interest $=$ principal $\times$ rate $\times$ time
Distance	distance $=$ rate $\times$ time

Order of Operations	The TI-30XS MultiView™ automatically evaluates numerical expressions using the Order of Operations based on how the expression is entered.	The correct answer is 23.

Example

$12 \div 2 \times 3 + 5 =$

Note that the 2 is **not** multiplied by the 3 before division occurs.

Decimals	To calculate with decimals, enter the whole number, then $\boxed{.}$, then the fractional part.	The correct answer is 17.016.

Example

$11.526 + 5.89 - 0.4 =$

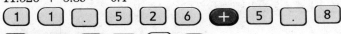

The decimal point helps line up the place value.

Fractions	To calculate with fractions, use the $\boxed{\frac{n}{d}}$ button. The answer will automatically be in its simplest form.	The correct answer is $\frac{15}{28}$.

Example

$\frac{3}{7} \div \frac{4}{5} =$

This key combination works if the calculator is in Classic mode or MathPrint™ mode.

Mixed Numbers	To calculate with mixed numbers, use the $\boxed{2nd}$ $\boxed{\frac{n}{d}}$ button. To see the fraction as an improper fraction, don't press the $\boxed{2nd}$ $\boxed{x10^n}$ buttons in sequence below.	The correct answer is $39\frac{13}{15}$.

Example

$8\frac{2}{3} \times 4\frac{3}{5} =$

This key combination only works if the calculator is in MathPrint™ mode.

Percentages	To calculate with percentages, enter the percent number, then $\boxed{2nd}$ $\boxed{(}$.	The correct answer is 360.

Example

$72\% \times 500 =$

$\boxed{7}$ $\boxed{2}$ $\boxed{2nd}$ $\boxed{(}$ $\boxed{\times}$ $\boxed{5}$ $\boxed{0}$ $\boxed{0}$ $\boxed{enter}$

Powers & Roots	To calculate with powers and roots, use the (x²) and (^) buttons for powers and the (2nd) (x²) and (2nd) (^) buttons for roots.	

To calculate with powers and roots, use the (x²) and (^) buttons for powers and the (2nd) (x²) and (2nd) (^) buttons for roots.

Example
$21^2 =$

[2] [1] (x²) (enter)

The correct answer is 441.

Example
$2^8 =$

[2] (^) [8] (enter)

The correct answer is 256.

Example
$\sqrt{729} =$

(2nd) (x²) [7] [2] [9] (enter)

The correct answer is 27.

Example
$\sqrt[5]{16807} =$

[5] (2nd) (^) [1] [6] [8] [0] [7] (enter)

The correct answer is 7.

You can use the (2nd) (x²) and (2nd) (^) buttons to also compute squares and square roots.

Scientific Notation

To calculate in scientific notation, use the (×10ⁿ) button as well as make sure your calculator is in Scientific notation in the (mode) menu.

The correct answer is 1.2011×10^5.

Example
$6.81 \times 10^4 + 5.201 \times 10^4 =$

[6] [.] [8] [1] (×10ⁿ) [4] (enter) (+)

[5] [.] [2] [0] [1] (×10ⁿ) [4] (enter)

When you are done using scientific notation, make sure to change back to Normal in the (mode) menu.

Toggle

In MathPrint™ mode, you can use the toggle button (◄►) to switch back and forth from exact answers (fractions, roots, π, etc.) and decimal approximations.

The correct answer is 0.428571429.

Example
$\frac{3}{7} =$

[3] (□/□) [7] (enter) (◄►)

If an exact answer is not required, you can press the toggle button (◄►) immediately to get a decimal approximation from an exact answer without reentering the expression.